ECO-TYRANNY

How the Left's Green Agenda Will Dismantle America

BRIAN SUSSMAN

Bestselling Author of *Climategate*

ECO-TYRANNY

WND Books
Washington, D.C.

Book Designed by Mark Karis

WND Books are distributed to the trade by:
Midpoint Trade Books
27 West 20th Street, Suite 1102
New York, NY 10011

WND Books are available at special discounts for bulk purchases. WND Books, Inc., also publishes books in electronic formats. For more information call (541) 474-1776 or visit www.wndbooks.com.

First Edition

ISBN 13 Digit: 978-1-936488-50-6

Library of Congress information available.

Printed in the United States of America.

DEDICATION

For Elisa, Sam, Ben, Josh, Makayla, and the next generation of patriots.
That you may able to stand your ground.

TABLE OF CONTENTS

PART TWO

FOREWORD

WHILE I CAN'T LAY CLAIM to being a lifelong conservative, I can declare that I've always been vehemently opposed to communism and socialism. No doubt, my parents influenced my convictions, particularly the story my dad told regarding his father's coming to America from Russia, shortly after the start Russian Revolution in the early 1900s. My grandfather was only thirteen years-old when his parents sent him to the United States with a single admonition: "Make money, send for your brother."

Indeed, Grandpa did just that. Without knowledge of the language, or the culture, he immediately embraced America's free-market system, buying used goods—a.k.a. "junk"—and reselling the merchandise door-to-door. Eventually, he sent for his brother, and they became wildly successful entrepreneurs.

Had my grandfather remained in Russia, the pall of communism would have darkened his life, as it did for millions of others. Instead, he came to America, eagerly partook of life, liberty, and the pursuit of happiness, and became a self-sufficient, contributing member of society. He was proud to be an American.

My grandfather's experience was just as the founders of this nation had intended. He was able to utilize his personal abilities to freely and independently engage in useful commerce. He would eventually purchase a home, construct a store (which he also owned), provide for his family, and create meaningful jobs for other citizens. Like millions of others, Grandpa was able to succeed because the government stayed out of his way, allowing him the opportunity to either make it, or fail, on his own accord.

My grandfather's success led to the overriding political philosophy espoused by my parents throughout my upbringing: "Government is but a necessary evil." Of course, this quote was actually extracted from Thomas Paine's pamphlet, *Common Sense*, published in 1776, in which Paine states, "Government, even in its best state, is but a necessary evil; in its worst state, an intolerable one."

What Grandpa escaped was the intolerable ideology of Karl Marx, which was sweeping his homeland when his parents put him on a ship bound for Ellis Island, New York. The core of Marx's message was the denial of private property. What Grandpa found in America was the antithesis of Marxism—the pursuit of happiness via his right to personal property, *and* a property in his personal rights.

The term "property," as understood by America's founders, does not limit the meaning to merely land or material possessions. Property includes one's thoughts, opinions, beliefs, ideas, and unalienable rights. No one articulated this ideology better than James Madison who, in a 1792 essay entitled, *Property*, stated:

> [Property] embraces every thing to which a man may attach a value and have a right; and *which leaves to every one else the like advantage.* In the former sense, a man's land, or merchandize, or money is called his property. In the latter sense, a man has property in his opinions and the free communication of them. He has a property of peculiar value in his religious opinions, and in the profession and practice dictated by them. He has property very dear to him in the safety and liberty of his person. He has an equal property in the free use of his faculties and free choice of the objects on which to employ them. In a word, as a man is said to have a right to his property, he may be equally said to have a property in his rights.

While much could be written regarding the usurping of property rights involving the non-physical, this book is dedicated to the tyranny that befalls the United States in the form of the physical—specifically land and natural resources. Even more so, this book exposes how environmentally-based government policies are being used as the lever of choice to engage in this travesty.

For many decades, the U.S. federal government has been instituting laws and regulations purposefully designed to smother physical property rights. Simultaneously, the feds have been engaged in amassing millions of acres of land—exactly what the founders *did not* want to occur. What the government is doing is part of a centralization of power, which Thomas Jefferson labeled the "feudal-ruler" form of government. In the feudal-ruler scheme, the rights of the government become superior to the rights of citizens. Such a system is foundational to Marxism in that the government demands control of property, because he who controls the land and its resources, controls the people. Thus, when citizens are unable to own land and property to grow crops, establish a business, or build a home to shelter their family, they are at the mercy of those who hold that power. Likewise, when a country's natural resources, necessary to sustain life, are withheld from the populace, or doled-out like airline peanuts, the citizens are beholden to the control of those in authority above them.

In other words, when the government controls the land, the water, the minerals, timber, oil, and gas, they control *us*.

ECO-TYRANNY

In 2010, WND Books published my first book, *Climategate, A Veteran Meteorologist Exposes the Global Warming Scam*. What makes that book unique is that in it I reveal the political and financial foundation driving those who are pushing the theory of anthropogenic global warming and climate change. However, while I was researching *Climategate*, I was astonished by a recurring theme: since the inception of the environmental movement, its leaders have been consumed with eliminating capitalism and ushering in a global era of socialism. Their call for being "green" goes far beyond demanding clean air, pure water, healthy forests, and alternative sources of energy. The leftists at the helm of the environmentalist hierarchy want to *control* the air, water, forests, and natural resources. Because I only touched on this research in *Climategate*, I felt it necessary to write another book that would provide the most comprehensive exposé of how the left's green agenda is trashing American liberty.

My interest in pursuing this book when into hyper-drive in 2010, when a former Department of Interior official in Washington, D.C. told me of a secret draft document created by the Obama administration that detailed a 25-year plan to purchase, or takeover, millions of acres of private property in order to connect land owned by, and managed by, the federal government. This colleague had seen a few pages of the paper, and believed it was part of a larger plot to "divide the country into sectors where all humans would be herded into urban hubs" with the bulk of the nation's land "returned to a natural state upon which humans would only be allowed to tread lightly."

The DOI official's words seemed to corroborate what my research revealed. After some digging, I was able to see the document for myself, at which point all of my suspicions were confirmed, and *Eco-Tyranny* was soon underway.

The plans that the left have for America are being tirelessly woven into the fabric of our society. Unless this plot is stopped, life for our children and grandchildren will not represent anything that once made America uniquely *American.*

FEDERAL LAND OWNERSHIP: ORIGINAL INTENT

Nothing is taught in our schools regarding federal land ownership. I'm sure most people think national parks and forests are as American as apple pie. Actually, they're not. The fact is our founders did not want the U.S. to be in the business of holding any property, other than that which is necessary to maintain a limited, central government.

So concerned were they about this issue that in October 1780, even prior to the ratification of the Articles of Confederation, the Continental Congress adopted a general policy for administering any North American land transferred *to* the Federal Government:

> The lands were to be "disposed of for the common benefit of the United States," and were to be "settled and formed into distinct republican States, which shall become members of the Federal Union, and shall have the same rights of sovereignty, freedom and independence, as the other States...." Additionally, the lands were to "be granted and settled at such times and under such regulations as shall hereafter be agreed on by the United States in Congress assembled..."[1]

The original intent was that any land bequeathed to the federal government, or obtained via treaty or purchase, would be allocated accordingly. This distribution resolution gave way to the original colonies ceding their "western" lands (between the Appalachian Mountains and the Mississippi River) to the central government between 1781 and 1802. In turn, ownership of these lands was transferred to private individuals and the states, with the proceeds being used to pay Revolutionary War soldiers, finance the new limited central government, and settle additional territories. Further federal land acquisitions continued thereafter, including the Louisiana Purchase in 1803, the Texas Annexation of 1845, the Oregon Treaty in 1846, the Mexican Cession of 1848, and the purchase of Alaska in 1867. However, the U.S. policy at the time of these arrangements was a planned transfer of ownership to individuals, states and territories; additionally government-owned land could also be used to house federal offices, create right-of-ways for interstate infrastructure (at the time, trains), and military installations.

Since our founding, the federal government has acquired 1.8 billion acres in North America, of which 1.1 billion acres have been properly disposed. The remaining lands are now under the control of four agencies: Bureau of Land Management, National Park Service, Fish and Wildlife Service (all divisions of the Department of the Interior), and the Forest Service, part of the Department of Agriculture. Specifically these gun-toting agencies control all public lands including national parks, national recreation areas, national forests, wilderness areas, heritage lands, wild and scenic rivers, historical sites, open spaces, and treasured landscapes.

United States' land policy markedly changed in 1832. It happened in Hot Springs, located in the Territory of Arkansas, so-named for its naturally-fed pools of hot spring water. The springs were considered therapeutic and able to cure a variety of ailments including arthritis, tuberculosis, skin diseases, and even cancer. Bleeding-heart members of Congress determined that the supposedly supernatural waters needed to be protected, and soon a first-of-a-kind law was made establishing the "Hot Springs Reservation." Next, a military hospital was constructed adjacent the springs to provide veterans the opportunity to be healed of various ailments by the miraculous waters. Eventually, some 100,000 hot baths were provided to the poor annually. Applicants for the free soaks were required to make an oath that they were without the means to pay for the baths, with violations being a misdemeanor subject to fine and/or imprisonment.

Hot Springs was the first slip down the proverbial, slippery slope.

The next slide came with the establishment of two iconic natural parks. In 1864, Congress donated the federally held Yosemite Valley to the State of California, with the proviso that the state would guarantee its preservation. While the decision was in keeping with the general intent of federal land ownership (transferring federal property to a state), it was a deal that was carried out with obvious strings attached: the feds were mandating what the state could do with the property. This land swap provided profound inspiration to an eclectic explorer, Dr. Ferdinand Hayden, of Massachusetts.

Hayden was a trained surgeon and geologist, as well as a self-described "naturalist."[2] He was also an incredible networker.

For many years, adventurers had spoken of the beautiful upper reaches of the Yellowstone River and a mysterious place nearby known as Colter's Hell. Located in what is now Wyoming, the region, filled with hot springs, steam vents, and geysers, was first discovered by a mountain man named John Colter. In 1804, Coulter headed west with the Lewis and Clark expedition, but separated from the group after a clash of personalities. That's when he happened upon the spectacular expanse near the Yellowstone River, which looked more like Hell than Heaven. Upon returning to civilization and sharing the stories of what he had witnessed, most cast him off as a kook.

Decades later, in 1871, intrigued by the stories put forward by Colter, Dr. Hayden led an impressive survey team to the Yellowstone Basin to find, and document, Colter's Hell. Upon his return East, Hayden mounted an emotional campaign to promote and protect the natural wonders of the region—just as had been done with Hot Springs in Arkansas, and Yosemite in California. Hayden made his case in a well-received article for a very popular publication, *Scribner's Monthly*. The piece included brilliantly detailed illustrations of the wonders his team had witnessed in upper Yellowstone, created by fellow expedition member Thomas Moran, one the nation's best painters. The public response to the article and realistic artwork were overwhelming.

Also on Hayden's journey was William Henry Jackson, arguably

America's best photographer. Following the piece in *Scribner's*, Hayden's next public relations blitz placed Jackson's stunning photographs on stereographic cards (postcards) produced by a friend, Charles Bierstadt, the leading manufacturer of the popular medium. Bierstadt fashioned the cards into a handsome album, and Hayden shared the photographs with Jay Cooke, a railroad magnate who helped finance the Union Army during the Civil War. Hayden was planning the Northern Pacific rail line and, as Hayden had hoped, Cooke became hopeful that putting Yellowstone on the map would increase ridership on his lines.

Cooke effectively lobbied Congress, passing around the photo album, and eventually Congress passed a bill calling for Yellowstone to officially become "a public park or pleasuring-ground for the benefit and enjoyment of the people."

The legislation was presented to President Ulysses Grant on March 1, 1872. Given the unique relationship between the former general of the Union troops and the Civil War financier, Mr. Cooke, there was little doubt that this bill would be vetoed. Once signed, Yellowstone became the first national park in the world.

With two major revisions to U.S. land policy completed, there was a third slip down the slope that has kept us sliding ever since: the Casa Grande National Monument in Arizona.

Casa Grande is an ancient village of mud huts and buildings created by indigenous people in what is now Coolidge, Arizona. The village was abandoned around 1450. As westward expansion increased, fanciful articles regarding Casa Grande's mysterious dwellings began circulating in the 1860s. Soon, with stagecoach travel available into the area, increased public interest was creating both a tourist attraction and a treasure hunt for collectors of antiquity.

Once again, a Massachusetts blueblood rose to the rescue.

In 1883, philanthropist Mary Hemenway financed an expedition led by anthropologist Frank H. Cushing to document the deterioration of the Casa Grande. Following the journey, Hemenway urged Massachusetts senator George F. Hoar to present a petition to Congress in 1889, requesting that the government take steps to repair and protect the ruins. The legislative efforts were successful, and restoration began the following year. In 1892, President Benjamin Harrison set aside one square mile of Arizona Territory surrounding the Casa Grande Ruins as the first prehistoric and cultural reserve established in the United States. Harrison derived his power to protect the property via the Land Revision Act of 1891. That act gave the president unilateral authority to "set aside and reserve...any part of the public lands wholly or partly covered with timber or undergrowth, whether of commercial value or not."

In other words, the act gave the president sole power to establish reserves, national forests, and national monuments. However, the law did not present the executive branch the authority to use or develop resources on the reserved lands.

By that time, our slip down the slope had become a slide gaining speed and out veering out of control. Terminal velocity was finally reached in the Taylor Grazing Act of 1934. Enacted to address the deterioration of public rangelands, this law provided direct authority for federal management of these lands, and implicitly began the aggressive move toward ending property disposals, and instead retaining land in federal ownership.

In 1976, with the Democrats in control of both bodies of Congress and Republican big-government advocates Gerald Ford and Nelson Rockefeller as president and vice president, the most sinister and greedy land grab law was crafted and signed. The Federal Land Policy and Management Act (FLPMA) officially declared that the U.S. was to retain the remaining lands in federal ownership and use the government's power to attain even more.

As you will later read, the 1976 FLPMA was everything eco-socialists desired, including the United Nations' environmental policy-maker, Maurice Strong. As you will see, that same year Strong would tell the World Conference on Human Settlements (known as "Habitat One"), "Public ownership of land is justified in favor of the common good, rather than to protect the interest of the already privileged."[3]

Perhaps the most tyrannous aspect of the FLPMA was the government's newfound ability to use eminent domain to confiscate and acquire private property for reasons heretofore never considered. In the past, the Constitution's Fifth Amendment provided eminent domain to be considered generally for the purpose of constructing infrastructure (roads, airports, reservoirs, federal buildings, etc.) that directly benefitted serving the needs of the people. The FLPMA now allowed the government to "acquire...by purchase, exchange, donation, or eminent domain, lands or interest therein," with the caveat that the properties "so acquired are confined to as narrow a corridor as is necessary to serve such purpose [in gaining access to] the National Forest System."[4]

Today, the FLPMA is used by environmental activists, who, strategically working with private individuals, non-profit organizations, and local and state governments, cobble together millions of acres land adjacent federally-owned property. Once acquired, the new parcels are given to the feds, thus forever taking the property off the books for future development.

The FLPMA, was succeeded by the Ecosystem Management Initiative (EMI), instituted by President Bill Clinton and Vice President Al Gore two decades later, as well as the founding of The National Landscape Conservation System (detailed later in Chapter Four). The EMI gave the federal government the power to pronounce and protect ecosystems (a very vague term), and the National Landscape Conservation System (NLCS) declared 27 million acres of federal land to be "landscapes" that must be protected. A primary goal of the FLPMA, EMI and NLCS is to continually hunt for more land to protect from future development, including the extraction of resources such as oil, gas, and coal.

President Obama is not without blame in this fall from liberty. As you will later see, his autocratic Great Outdoors Initiative was the most aggressive attempt to codify these elements of eco-tyranny.

NO WAY

Over the course of reading *Eco-Tyranny*, you will likely experience moments where you literally find yourself exclaiming aloud, "No way!"

My wife will be the first to tell you that throughout my media career in San Francisco, both as a television meteorologist and science reporter, and then as a radio talk show host, I've presented my audience with facts that commonly generate a "no way" response.

In this book, most of the "no ways" will be generated from facts and figures related to the environmental movement, and often you will be shocked.

However, this book is not intended to create hysteria; it's a book designed to expose the truth. I want Americans to see how Marxists have hijacked the environment in order to push a grossly anti-American agenda. They are using our schools, our religious heritage, non-profit institutions, businesses, and all facets of government to forward their radical plans.

But, there is hope. Besides a comprehensive investigation into their liberty-vanquishing plan, I will also present solutions on how we can restore our property rights, reclaim our resources, create jobs, and reassert ourselves as the most exceptional nation on the planet.

The timing of the release of *Eco-Tyranny* is also no accident. *Climategate* was published prior to a vote on the America Clean Energy and Security Act, a.k.a., Cap-and-Trade. My goal was to arm the voters with the facts so that the bill would not become law. The legislation passed the Nancy Pelosi-led House of Representatives, but it failed to make it through the Senate. That failure was a major victory. Many legislators who supported the legislation were voted out of office in the 2010 elections by informed, motivated, and activated constituents.

Now, this book arrives in time for one of the most important elections in decades. My hope is that it will serve to educate, motivate, and activate *you*, to make sure that representatives are elected into pubic office with the intention of reversing the leftist course of the wrongs that have been done, and placing us back on a path toward eco-liberty.

Let us be reminded of timely words of wisdom, spoken by an original patriot:

> No people will tamely surrender their Liberties, nor can any be easily subdued, when knowledge is diffused and Virtue is preserved. On the Contrary, when People are universally ignorant, and debauched in their Manners, they will sink under their own weight without the Aid of foreign Invaders.
>
> —Samuel Adams, signer to the Declaration of Independence

PART ONE

ONE

TYRANNY SPAWNED

POLLUTION NEVER HAS BEEN Earth's most troubling foe—Marxism has. And Marxists have always seized upon pollution, both real and imagined, as an effective weapon in their unrelenting war on freedom.

Karl Marx founded a philosophy that inspires dictators and demagogues. Commencing with the Russian Revolution in 1917 to the present, Marx's tyrannical ideology has been responsible for the documented deaths of more than 110 million individuals around the world. Hundreds of millions more have been forced to live in oppressed societies, void of the inalienable rights of life, liberty, and the pursuit of happiness.

Yet despite the suffering Marxism has unleashed on the planet, we are continually lectured—by politicians, government bureaucrats, professors, environmental groups, and movie stars—that the world's foremost enemy is pollution, particularly greenhouse gas emissions. Our greatest challenge, they insist, is curtailing such discharges into the atmosphere and restoring the global environment; if we don't, they claim, the earth's ecological system will die.

It's all a lie.

There is no such planetary crisis. It's a concocted calamity churned out initially by Marx himself, and furthered by his modern devotees. It's what I have named the "green agenda."

Is there pollution? Sure. Can it be cleaned up? Absolutely—and in the United States we have done a remarkable job of doing so. But to declare that there is a dire *global* eco-emergency, particularly one that is fueled by the use of fossil fuels and subsequent carbon dioxide (CO_2) emissions, emanates from an anticapitalist plot that's been playing for nearly 150 years.

Born in Germany in 1818, Karl Marx lived sixty-five years, during which time his twisted mind conceived and perfected an atrocious plot to infect the world with his philosophy of "organized collectivism"—a.k.a. communism, or, for the more politically correct, socialism. Marx began his rebellion as a student at the University of Berlin, where he was strongly influenced by the philosophy of radical thinker Georg Hegel. Hegel held that Christianity had a negative effect on society and that a new religion built on scientific reason was needed. Marx was so taken by Hegel that he joined a group known as the Young Hegelians. Their initial goal was straightforward: liquidate Christianity.

What was the Young Hegelians' beef with the Church? They were convinced it was a system of *beliefs*, as opposed to facts. Christianity proclaimed there was a single God who was said to be especially fond of the human race; this God even created heaven for those who are good and hell for those who are bad. In addition to these beliefs, Christians (and Jews—the Hegelians weren't keen on their religion either) believed that God made them in His image and told them they were to "fill the earth and subdue it; and have dominion over the fish of the sea and over the birds of the air and over every living thing that moves upon the earth" (Gen. 1:28). While conservative Bible scholars are convinced the original language in the verse implies reasonable stewardship of the earth's magnificent resources, Hegel saw this verse as giving religious zealots license to overpopulate the planet and rape its assets. Likewise, the Young Hegelians saw religion as a societal illness that could be cured through proper scientific reasoning and education.

Hegel also lectured that everything in the universe could be explained through his system of rational thinking, known as the "dialectical process." This idea was originally conceived by Greek philosophers; Hegel's version of this process held that contradictions in nature do not harm one another, but instead lead to a higher level of development, particularly personal development—hence, no need for religion. Marx would eventually take Hegel's dialectic theory and refine it for political purposes.

In 1841, Marx received a doctorate in philosophy; his thesis was on Epicurus, the ancient Greek philosopher and quasi-atheist who taught that the physical world was all there was and all there would ever be. Epicurus believed the fundamental constituents of the earth's system were indivisible, invisible bits of matter known as "atoms." By Marx's time, Epicurus's theory of matter was known as "materialism," and Marx was a true materialist.

MARX AND MATERIALISM

In 1842, Karl Marx met Frederick Engels, and together, the two began developing a propaganda campaign that would alter the course of the world. Holding parallel views on materialism, dialectics, and even the abolition of a supreme being, the two were convinced science was the ultimate path to godlike superiority. They held that "if science can get to know all there is to know about matter, we will then know all there is to know about everything."[1]

For Marx and Engels, matter—atoms, molecules, and the otherwise unseen—was the alpha and omega of reality. Matter provided the complete explanation for plants, animals, man, intelligence, planets, and solar systems. They also believed that time, certainly not a divine Creator, was the magic wand that allowed all matter to come together to create the universe in which we live. Marx and Engels's convictions were later neatly articulated by modern astronomer Carl Sagan, a prime guru of today's environmentalists, who most famously said, "The cosmos is all there is, or was, or ever will be."[2]

To codify their secular belief system, Marx and Engels prescribed three laws of matter: the law of opposites, the law of negation, and the law of transformation. Together, these planks provide the rationale for today's green agenda.

The Law of Opposites is an extension of Hegel's work and supposedly illustrates how everything in existence is a combination of dialectics working in unity. For example, electricity is characterized by a positive and negative charge. Atoms include protons and electrons, which are contradictory forces working in unity. Even the human race is composed of opposite qualities: altruism and selfishness, courage and cowardice, humility and pride, masculinity and femininity. To function properly, Marx believed, these opposite forces must be kept in balance; if they aren't, discord is certain. Thus, the law of opposites demands that humans must be kept in check, because, as the most advanced creatures, they can wreak the most havoc—hence, the need for a tightly regulated, often heavy-handed system of government. This is why socialists in the United States slobber over Castro's Cuba and herald Venezuela's Hugo Chávez, as heroes. According to the law of opposites, demagogues are essential to effective, masterful governance.

The law of negation adds a somewhat metaphysical component to Marx's madness that provides a key pillar for today's environmentalists, declaring that all nature is constantly expanding through death. To support this "law" Engels created an awkward illustration using barley seed, which germinates—via its own death or negation—and produces many new plants. This concept seems especially clumsy when applied to the human race, but Engels rationalized it by claiming that "out of this dynamic process of dying the energy is released to expand and produce many more entities of the same kind."[3]

In other words, all species possess an inherent tendency to proliferate. However, Marx and Engels believed nonhuman species bear automatic mechanisms to properly manage such expansion and prevent their increases from growing out of control, but the extended family belonging to Homosapiens are incapable of such self-regulation. Thus, the law of negation casts mankind as an ever-consuming population bomb that places the entire planet at risk. As a result, negation insists that systems must be put in place to maintain sustainability, including mechanisms to ensure human population control when necessary.

The law of negation is the motivating factor for an eco-buzzword with which we are constantly assailed today: "sustainable." Toyota tells us its Prius is "the vehicle that started the sustainable transportation revolution."[4] Whole Foods Market states that they sell fish that were "caught in an environmentally sustainable manner."[5] The intentions of such marketing plans are clever, but carried to an extreme, the concept of sustainability involves much more than battery-assisted cars and farmed fish—it's a call for government policies that demand changes in human behavior and lifestyles under penalty of law.

The third Marx/Engels axiom is the most arrogant: the law of transformation, which states that a continuous quantitative development by a particular species often results in a "leap" within nature, whereby a completely new form or entity is produced. This law was bolstered by the findings of Marx and Engels's contemporary, Charles Darwin. Darwin's theory of evolution sealed the communist founders' convictions that such "leaps" not only allowed for the origin of new species, but a leap *within* a species—particularly Homosapiens—which could enable some to advance to new levels of reality.

Thus, the law of transformation confirmed an elite status within the human race; and those born into evolution's aristocracy possess a duty to dictate how the underdeveloped shall live. Taken to an extreme, transformation could also determine who shall die.

With these new revelations, Marx and Engels arrogantly boasted, "The last vestige of a Creator external to the world is obliterated."[6]

I've often summarized the laws of matter this way:

> Committed Marxists are convinced that phenomena such as love, passion, value and feelings aren't "real" because they're not composed of matter; even consciousness, and especially faith in God, are simply the result of material interactions within the human mind. In addition, Marxists contend that some people are randomly spit out of their mother's womb with a better brain than most. Those with the best brains have a Darwinian authority to rule over those with the lesser brains, lest those with the deficient brains destroy the planet and kill one another; thus, the need for a heavy-handed form of government loaded with burdensome regulations, and the perfect excuse for socialism, communism, and fascism.

Over the ensuing years, Marx never drifted from his materialistic assumptions and antagonist view of Christianity. Instead, he was able to neatly tuck those ingredients into his theory of organized collectivism, providing it with a holistic framework from which society could be rebirthed.

Today, such a patrician worldview resonates with those who consider themselves to be politically liberal, and resonates even more so with those liberals who are the direct products of America's elite colleges and universities. Through advanced education and absorption of information, this "elite" caste surmises that they can become masters of the universe. Theirs is an amoral system in which there is no room for absolute truth, only relativism, lest they be forced to acknowledge a divine being who has an absolute rule of law, which would force them to throw out their dogma, or else move forward filled with guilt about their waywardness.

Marxists believe *they* have the power to define all societal morality, rules, and laws subject to their goals. Hence, the inalienable rights of life, liberty, and the pursuit of happiness as recognized by America's founders are viewed as absurd, because an imaginary God cannot declare rights. Marxism demands that all so-called rights be issued by the government in the form of laws. And just as a law can be issued by the government, so shall it be taken away by that government if deemed necessary.

WHY MARX LOATHED AMERICA

Karl Marx and his followers were well aware of the Republican form of government established in America—and loathed it. Marx perceived America's founders as reckless, religious buffoons who were peddling dangerous propaganda—especially the inalienable rights of their Declaration of Independence. To the collectivist, such freedoms were—and are—preposterous. The *life* of an individual is not unique—just a fragment of the ever-multiplying collective mass—the result of a random, cosmic, Darwinian accident. Likewise, *liberty* is an unattainable sentiment. Left to their own devices, the human masses are wholly incapable of coexistence without formidable government control and regulation. Furthermore, the *pursuit of happiness* is the most egregious maxim of all, and Marx was aware of the origins of this key phrase. It was penned in direct reference to the words of English philosopher John Locke, who in his 1690 essay, *Concerning Human Understanding* wrote, "The necessity of pursuing happiness [is] the foundation of liberty."

Locke also made the tenacious argument (in chapter 5 of the *Second Treatise* on civil government, from his work *Two Treatises of Government*) that human happiness is directly linked to one's personal property, and that property included tangible elements, such as land, natural resources, and material goods, and, as well, the intangible elements of speech, thoughts, and beliefs. Locke further stated that God had even given mankind the right to physically defend his property.

America's founders understood that property was synonymous with liberty and security. They comprehended that in a capitalistic, free-market economic system void of overreaching, central government regulation, a new worker or immigrant could progress up the class ladder in conjunction with his or her effort and reap the happiness associated with owning his or her own business, farm, home, and estate. Regarding the potential for abuse within such a capitalistic framework, George Washington insisted that "virtue or morality is a necessary spring of popular government."[7]

Likewise, John Adams said, "Public virtue cannot exist in a Nation without private Virtue, and public Virtue is the only Foundation of Republics."[8] In other words, our Founders realized there would be scoundrels who would abuse American liberty, but they knew such risks are worth taking, as the alternative would lead down the slippery slope of government-led tyranny. As Patrick Henry famously proclaimed in 1775: "Is life so dear, or peace so sweet, as to be purchased at the price of chains and slavery? Forbid it, Almighty God! I know not what course others may take; but as for me, give me liberty or give me death!"[9]

Such professions were highly offensive to Marx, as he was convinced that virtue and morality were the products of confused brain matter. Capitalism, he believed, *always* results in a class struggle between the owners and controllers of the means of production (the *bourgeoisie*) and the mass of common laborers (the *proletariat*). Marx was convinced the proletariat were oppressed and exploited by the bourgeoisie—period. He also concluded that because capitalism is designed to seek a profit, it is the proletariats who suffer, because profit is made off of their labor, and their labor causes the degeneration of their lives, health, and ability to enjoy life.

As an antidote to the presumptuous experiment being conducted in the United States, in 1849, Marx and Engels presented to the world their final formula for revolution, which they called the *Manifesto of the Communist Party*. This infernal document would eventually be known as *The Communist Manifesto*.

In chapter 2 of their manifesto, Marx and Engels boldly stated the goal of their envisioned new world order: "The theory of the Communists may be summed up in the single sentence: abolition of private property."

Property is not just the house and land you may own—it's your car, possessions, business, thoughts, ideas, beliefs, and goals. It was as if Marx and Engels looked both God and humankind in the eye, raised their middle fingers and shouted, "No!" Personal property, they declared, is a myth; *matter* is all there is, and no individual can claim matter as his or her own.

GUANO: THE FIRST ENVIRONMENTAL CRISIS

Marx perceived the environment as an effective apparatus to justify his hatred of capitalism and liberty, though at the time, the term used was not "environment" but "nature." In many ways, Karl Marx was the premier environmentalist and founding father of the green agenda.

In 1862, a fellow materialist and academic colleague of Marx, German chemist Justus von Liebig, published an updated version of an otherwise boring book he wrote twenty-two years prior, titled *Organic Chemistry in Its Application to Agriculture and Physiology*. The new edition was unique in that for the first time, a scientist used his lectern to create an environmental argument to attack capitalism. The issue at hand was bird droppings, or *guano*.

In the mid-1800s, Britain's citizens were living longer and healthier lives in comparison to the rest of the world. Much of this good fortune was the result of newly developed domestic farming techniques that were able to deliver an abundance of affordable food to the people. One of the key ingredients to the Brits' farming success was the use of guano, an efficient fertilizer. Farmers were willing to purchase the ordure from anyone who would sell it at a reasonable price. Guano imports to England first began in 1841, and twenty years later, it is estimated some 3.2 million tons of the phosphate-rich additive had been brought into the country.[10] Guano was being carried to market from mountaintops, fields, and caves in Europe, North America, South America, Africa, and the Caribbean islands.

Though von Liebig well understood the theoretical benefits of utilizing guano as a fertilizer, he had significant personal displeasures regarding its usage. First, he felt that guano hunters were destroying nature while collecting deposits of the organic material. Second, he contended that greedy guano traders were taking advantage of underpaid workers in order to turn a profit. Third, von Liebig was angered that the crops that benefited from the guano were growing at a rate that he believed superseded nature's intention. This increase in yields created more feed for livestock, and more vegetables and meat for Londoners to eat. The people were now living longer, healthier lives and tended to have larger families. Larger families, von Liebig complained, require larger houses and more animals for transportation. More food was necessary to sustain these larger families. Lastly, a teeming population and additional animals meant more excrement and pollution.

Von Liebig described guano as being at the center of a "robbery system." Using variegated imagery, von Liebig said Great Britain's use of guano

> deprives all countries of the conditions of their fertility. It has raked up the battlefields of Leipsic, Waterloo and the Crimea; it has consumed the bones of many generations accumulated in the catacombs of Sicily; and now annually destroys the food for a future generation of three millions and a half of people. Like a vampire it hangs on the breast of Europe, and even the world, sucking its lifeblood without any real necessity or permanent gain for itself.[11]

Dr. Justus von Liebig was the first recorded fellow traveler with a physical science PhD to attack capitalism based on environmental standards. His strategy would be thoroughly vetted by Marx and eventually become the lever of choice for future socialists and communists in the twenty-first century who, like a henhouse full of Chicken Littles, would try to convince the world that the earth is warming, the ice is melting, the sea is rising, the polar bears are dying—and it's all *your* fault.

MARX GOES GREEN

At the time of von Liebig's launch on Great Britain's agricultural methods, Marx was in the process of completing one of his signature works, *Das Kapital.* Marx was quite impacted by von Liebig's complex book on organic chemistry and in an 1866 letter to Engels, wrote, "I had to plough through the new agricultural chemistry in Germany, in particular von Liebig...which is more important for this matter than all of the economists put together."[12]

Marx believed that the earth possessed what he commonly referred to as "natural wealth," which he described in *Das Kapital* as "fruitful soil, waters teeming with fish, etc., and...waterfalls, navigable rivers, wood, metal, coal, etc."[13] Marx, like von Liebig, was convinced such natural resources did not belong to man and could only be utilized if necessary for the absolute common good, and without anyone garnishing a profit along the way.

In *Das Kapital*, Marx went on to state that one of "von Liebig's immortal merits" was having "developed from the point of view of natural science, the negative, i.e., destructive, side of modern agriculture."[14]

Regarding von Liebig's extreme criticism of the guano trade, Marx focused less on the cycle of secretions and more on the sheer economics. Marx felt that in addition to British farmers exploiting natural wealth (in this case, the natural wealth was twofold: the guano and the crops) in order to garner a surplus profit, the upper class was exploiting the lower class.

He saw it like this: lowly human laborers, who were used to extract guano in faraway places for shipment to England, were being sorely underpaid and thus abused by their well-to-do employers. Marx perceived further injustice as the guano was spread on the fields by low-wage laborers.

As the crop yields expanded, natural wealth was further abused as the farmers gained higher profit margins. The bountiful crops also provided an overabundance of feed for the livestock, which enabled their upper-class owners to reap additional profits by raising more animals for less.

Like von Liebig, Marx also believed that eventually city dwellers were able to purchase more food at a lesser cost, encouraging them to have larger families. Larger families required bigger houses to be built by more exploited workers. The urban population boom required more horses for transportation and more subsequent waste (animal dung and the human stuff too) to be removed from the cities and hauled to the dump, again by exploited, underpaid laborers.

Marx perceived all of this as nothing more than a vicious cycle created and perpetuated by a lust for profit. He succinctly described the process of this modern system of agriculture, stating, "The increased exploitation of natural wealth by the mere increase in the tension of labor-power, science and technology give capital a power of expansion."[15]

Further, Marx implied that this new capital-based system of agriculture was irrational. "The moral of history," he said, "is that the capitalist system works against a rational agriculture, [and] that a rational agriculture is incompatible with the capitalist system."[16]

What Marx stated in *Das Kapital* is still held as gospel by his minions today: no one has the right to make a profit off of natural resources, such as food, water, timber, coal, gas, or oil. And whether it's saving the forests, whales, snails, or the climate, it all comes back to a deep-rooted belief that the quest for such profit is immoral and will ultimately destroy the planet unless ground to a halt.

Conversely, as a conservative looking upon British farming in the 1860s through the rearview mirror of history, I see an excellent market opportunity. Farmers recognized a demand for their vital products and wished to increase the supply. Science proved that by adding nutrients to the soil, crop yields would become robust. Bird dung, a messy substance no one was fond of, was discovered to be rich in phosphate nutrients, and farmers began using it on their fields with great success. It was a win-win: the droppings actually had value, and farmers began collecting the poop wherever it was found. When the local pool of droppings was expended, businesses were created to locate and import new sources of the wonderful fertilizer. Jobs were created as people were hired to work in all phases of the guano trade, and their lives were made better. Ships were built and various tools were fashioned to support the new industry—again employing more people.

In the farming community, business was booming. More crops meant lower prices for the consumer and higher profit margins for the grower. Feed for raising cattle, sheep, pigs, and chickens was more abundant. For the first time, food rich in protein was available to more people, and their health was subsequently improved. Additional businesses were established to support the farming industry. Britain's unemployment rate decreased, crime was reduced, and lifestyles vastly improved.

Yes, there was more human and animal waste, but its cleanup also created a new industry—waste management. It, too, provided both honest work and the incentive for some to get better jobs.

However, Marx was incapable of envisioning prosperity in a positive light, and his followers are beset with that same problem today. In fact, in *Das Kapital*, Marx sounds like a modern-day environmental activist:

> All progress in capitalistic agriculture is a progress in the art, not only of robbing the laborer, but of robbing the soil; all progress in increasing the fertility of the soil for a given time, is a progress towards ruining the lasting sources of that fertility. The more a country starts its development on the foundation of modern industry, like the United States, for example, the more rapid is this process of destruction.[17]

Today's followers of Marx have not changed. They continue to perceive capitalism as unjust, the use of natural resources for profit immoral, and the human population something that must be controlled.

FROM MARX TO LENIN

There are three additional founders of the green agenda that need to be brought into our discussion, whose names you may be unfamiliar with, but who certainly need to be noted, as they are revered by environmentalist teachers and leaders today.

Sir Edwin Ray Lankester was a zoologist at University College, London, and noted as the greatest Darwinist of his generation—in fact, it is well established that Lankester's family were friends with Charles Darwin, and much has been written of little Ray being "carried on the shoulders of Darwin" as a child.[18] Though Lankester was some thirty years younger than Marx, the two were close friends, colleagues, fellow materialists, and socialists. Lankester was a frequent guest at Marx's household in the last few years of Marx's life and attended his funeral.

Regarding *Das Kapital*, Lankester once wrote Marx that he was absorbing Marx's "great work on *Kapital*...with the greatest pleasure and profit."[19]

Lankester was the most eco-socialistic thinker of his time, writing powerful papers on species extinction due to human causes with an urgency that would not be not found again until the late twentieth century. Lankester's most popular screed was *Nature and Man*, in which he described humans as the "insurgent son[s]" of Nature.[20] According to Lankester,

> We may indeed compare civilized man to a successful rebel against nature who by every step forward renders himself liable to greater and greater penalties... He has willingly abrogated, in many important

> respects, the laws of his mother Nature by which the kingdom was hitherto governed; he has gained some power and advantage by so doing, but is threatened on every hand by dangers and disasters hitherto restrained: no retreat is possible—his only hope is to control...the sources of these dangers and disasters.[21]

Lankester's star pupil was Arthur Tansley—the man noted for coining the term *ecosystem*. Born in 1871, Tansley was never able to interface with Marx, but was a fellow Darwinist, materialist, socialist, and a foremost academician specializing in botany. Tansley was deeply concerned with "the destructive human activities of the modern world." He argued, "Ecology must be applied to conditions brought about by human activity."[22]

In the 1940s, Tansley had a young protégé named Charles Elton, who worked with him to further develop the ecosystem concept. Elton's fiery writing style set the stage for the coming generation of eco-authors. In a blazing 1958 condemnation of the use of pesticides, Elton declared that "this astonishing rain of death upon so much of the world's surface" was largely unnecessary and threatened "the very delicately organized interlocking system of populations" in the ecosystem.[23]

From Karl Marx to Charles Elton, a mere three degrees of separation bring us to the modern, radical environmental movement. However, there is another historical figure that must be properly highlighted, in that he was the first political leader to practically implement the green agenda.

VLADIMIR LENIN: THE COMMUNIST GREEN GIANT

Vladimir Ilyich Lenin. Mention the name to any U.S. citizen who is formerly from the Soviet Union and the instant response will be visceral. Lenin was the Marx-honoring Communist who overthrew Russia and birthed a movement of tyranny that eventually plunged Russia and eastern Europe into several generations of doom and misery.

Lenin was born in 1870 into a family steeped in revolutionary thought. When he was seventeen, his older brother was executed for attempting to assassinate the czar. Several years later, Lenin began to engross himself in the works of Marx. By the early 1900s, he was a well-known Marxist author writing books on materialism and socialist economic theory. In 1916, he penned the angry missive *Imperialism, the Highest Stage of*

Capitalism. Having gained a significant following by October of the following year, he and a small band of cohorts staged a cunning coup. Lenin was named chairman of the new government, and the Russian Revolution had begun.

Immediately, members of the former regime were arrested and in many cases immediately executed. Banks were quickly nationalized, private businesses taken over by the state, and a supreme economic council was formed to run the economy. All private land, including any property belonging to the church, now belonged to the new Soviet state. A civil war ensued as freedom lovers tried to withstand the new government and its Red Army, but they were eventually, and brutally, defeated. Estimates vary as to the number who died in the war, but between men killed in action, those executed by the Red Army, civilian casualties, and military members who perished of exposure and disease, the number is likely more than a million. Russia's economy was devastated by the war, with factories and infrastructure destroyed, livestock and raw materials pillaged, mines flooded, and the people without hope.

Lenin's country was a horrific mess. Yet one of his top priorities from the very beginning of his brief seven-year reign was to institute a green agenda. Besides being a devout student of Marx, Lenin was familiar with von Liebig and Lankester and, like them, believed that nature's resources should never be used for profit—only for the good of the people, and then, only if absolutely necessary. In fact, as the Supreme Soviet leader, Lenin would be known for being far kinder to nature than he was to the people he ruled.

Within his first year as party chairman, 1918, he issued a mandate titled Decree on Land. It declared all forests, waters, and minerals to be the property of the state. Later that same year, as locals began to clear portions of the forest for firewood and construction material, Lenin issued a stern diktat titled Decree on Forests. From that moment on, the forests were protected and only certain small, insignificant sectors were established for harvest. Lenin's decree declared the protected areas as a "preservation of monuments of nature."[24]

Animal rights came next with the decree "On Hunting Seasons and the Right to Possess Hunting Weapons," which began to be enforced in May 1919. It banned the hunting of moose and wild goats, and ended the open seasons for a variety of other animals in spring and summer.[25]

Lenin's counsel in crafting this green agenda came from acclaimed Russian agronomist N. N. Podiapolski, who urged the creation of *zapovedniki,* or nature preserves. In such preserves, nature would be totally left alone—no hunting, harvesting, clearing of dead growth, mowing, sowing, or even the gathering of fruit. Humans would not be allowed in such regions. Podiapolski recalled one meeting with Lenin, convened despite the chairman's involvement in a fierce military campaign:

> Having asked me some questions about the military and political situation in the Astrakhan region [a fertile area rich in natural resources located in southwestern Russia on the delta of the Volga River, sixty miles from the Caspian Sea], Vladimir Ilyich expressed his approval for all of our initiatives and in particular the one concerning the project for the zapovednik. He stated that the cause of conservation was important not only for the Astrakhan region, but for the whole republic as well.[26]

Podiapolski drafted a resolution that was eventually accepted by the Soviet government in September 1921, titled "On the Protection of Nature, Gardens, and Parks." A commission was established to supervise execution of the new laws. One of its first tasks was to create another zapovedniki (named the Ilmensky preserve) in the Chelyabinsk region on the slopes of the south Ural Mountains, an area rich in coal, iron ore, nonferrous metals, and gold. Despite the great potential economic value to the state, Lenin believed the minerals were much more important for what they could teach scientists about geological processes. In Lenin's Russia, scientific understanding took priority over meeting the needs of his people, the majority of whom were living in utter poverty.

You must understand, the green agenda—from guano to global warming—is not about celebrating the beauty of our planet; it is an assault on mankind. It's an agenda that has no regard for your needs, lifestyle, dreams, desires, or feelings.

During Lenin's reign, Russia initiated the most audacious nature conservancy program in the twentieth century. Starting with a vision created by Marx fifty years prior, Lenin had successfully implemented version one of the green agenda. His accomplishments would eventually (as you will soon see) be celebrated the world over each April.

HITLER'S VERSION

Nazism was another type of socialism practiced by the National Socialist German Workers' Party (aka the Nazi Party), led by Adolf Hitler in the 1930s. Hitler's Nazis held many key tenets in common with the philosophy articulated by Marx. They believed in a government-regulated, planned economy. More important, though Hitler was agreeable to the limited private ownership of property, he was against capitalism and those who sought undo personal profit. During his reign of terror, a popular government slogan was "Fixing of profits, not their suppression."

Hitler also believed capitalism was created by the Jews and once told Italian Fascist dictator Mussolini, "Capitalism has run its course."[27]

So determined was Hitler to eliminate excessive profit that execution or imprisonment in a concentration camp was the punishment exacted for any business owner who pursued his own self-interest, instead of the interests of the state. This official decree was stamped into the rim of the silver reichsmark coins issued during the Nazi reign: "*Gemeinnutz geht vor Eigennutz*," or "The common good before self-interest."

Additionally, like Marx, Hitler saw nature, not humankind, as supreme. He also insisted that the environment be protected from commercial development, stating, "Man must not fall into the error of thinking that he was ever meant to become lord and master of Nature."[28]

Like his communist cousins, Hitler perceived pollution as the direct result of capitalism and, as a fellow materialist who spurned belief in the supernatural (despite occasional vague rhetoric about "Providence"), he was completely opposed to the harvesting or mining of natural resources, stating, "The German countryside must be preserved under all circumstances, for it is and has forever been the source of strength and greatness of our people."[29]

Following Marx, Lenin, and Hitler, a new generation would go forth to further unfurl socialism's green flag, and they would achieve great success with their devious agenda...even in America.

TWO

GREEN THE NEW RED

ENVIRONMENTALIST ACTIVISTS ARE DOGMATIC, ideological radicals hell-bent on transforming society into a colossal, highly regulated, redistributive commune void of inalienable rights. Their lack of integrity enables them to look you straight in the eye and lie about the facts, while they spin out tailor-made, cherry-picked research supposedly proving their many fictitious claims regarding the state of the global ecosystem. The primary goal of their green agenda is not a pristine environment—it's about gaining absolute control over your life. A contemporary example of this desire was expressed by President Barack Obama's former "Green Czar," Van Jones—a self-described communist—who stood before a rally attended by thousands of young eco-radicals in 2009, shouting, "This movement is deeper than a solar panel, deeper than a solar panel. Don't stop there. Don't stop there. No, we're going to change the whole system. We're going to change the whole thing!"[1]

While Charles Elton's writings from the 1950s successfully illustrated that ecological issues could be used to smear capitalism and American liberty, stronger arguments were interjected into the popular culture in the sixties. Two cunning authors, Rachel Carson and Paul Ehrlich, independently concocted denunciations that cleverly mixed ecology with sociology and political science. Though their books were best sellers, they were peppered with statistics and statements that could not withstand the microscope of truth. Nevertheless the authors, perhaps knowing that most readers would not have the patience or desire to cross-reference their work, brazenly stated what they determined necessary in order to cast our republican form of government and its free markets as failed experiments that could only be resolved with progressive doses of socialism.

Carson came first. She held a degree in zoology and found employment creating brochures for the United States Department of Fish and Wildlife from 1936 to 1950. Though Carson longed to become a successful author, her first effort fell flat, but her second book, *The Sea Around Us*, proved her to be a gifted word stylist. Published in 1951, *The Sea Around Us* is a flowery ode to the mysteries and magic of the sea, within which scientific information is seamlessly interwoven. A decade later, with the release of her epic *Silent Spring* in 1962 (which loses the entrancing expressions found in her first book and instead reads like a breathless, 363-page assault on mankind), Carson revealed how her beliefs were influenced by a cadre of Marxists.

One such associate was H. J. Muller, a Nobel Prize recipient in genetics. Throughout the pages of *Silent Spring*, Carson made several references to Muller's academic work regarding radiation. However, she failed to mention that Muller was an anti-American communist, whose biased research on radiation was conducted to make a case that the United States nuclear arsenal must be destroyed.

In the early 1930s, as a professor at the University of Texas, Muller became a faculty adviser to the National Student League (NSL), a well-known communist organization that was present on many American college and university campuses. At one point, NSL members pledged, "We will not support the government of the United States in any war it may conduct."[2] Muller even helped sponsor and edit the club's publication, *Spark*, named after Vladimir Lenin's newspaper, *Iskra*, which means "spark."[3] Muller eventually became so disenchanted with America that he moved to Nazi Germany in 1932 and then eventually to the Soviet Union.

Muller was one of Carson's go-to experts in providing "research" for *Silent Spring*, which essentially made her a covert Soviet plant.

Carson was also taken by the work of previously noted British ecologist Sir Arthur George Tansley, who was mentored by friend-of-Marx Edwin Ray Lankester. Tansley was such a cosmic thinker that he once wondered if humanity "is a part of nature or not."[4]

Tansley's mentee, British zoologist Charles Elton, was fundamental to the development of Carson's ecological critique. Elton wrote *The Ecology of Invasions by Animals and Plants* (1958), in which he employed the new ecosystem concept that was to inspire much of the wider arguments presented in *Silent Spring*. In a powerful ecological condemnation of synthetic pesticides, Elton declared that "this astonishing rain of death upon so much of the world's surface" was largely unnecessary and threatened "the very delicately organized interlocking system of populations" in a given ecosystem.[5]

Carson quoted Elton's "rain of death" line in an April 1959 letter to the *New York Times* in which she introduced her attack on pesticides; she quoted it again in *Silent Spring* in the chapter "Indiscriminately from the Skies."[6]

A fourth significant influence on Carson was her friend Robert Rudd, a professor of zoology at the University of California at Davis. Rudd has been described by a noted socialist historian as "a sophisticated left thinker with a deep sense of the ecology, sociology, and political economy."[7] Carson first contacted Rudd in April 1958 to receive assistance in writing *Silent Spring*. Carson and Rudd are said to have developed a strong friendship and a close working relationship.[8] Carson drew extensively on Rudd's research in two of *Silent Spring*'s chapters, morosely titled "And No Birds Sang" and "Rivers of Death."

For Carson, ecology had emerged as the basis for a radical challenge to the human domination of nature. "The modern world," she declared, "worships the gods of speed and quantity, and of the quick and easy profit, and out of this idolatry, monstrous evils have arisen."[9] In a clarion call for the formation of a global environmental movement to combat modern development, Carson proclaimed that "the struggle against the massed might of industry is too big for one or two individuals...to handle." And borrowing directly from Marx's *Laws of Matter*, Carson stated in a rare television interview that "man's endeavors to control nature by his powers to alter and to destroy would inevitably evolve into a war against himself, a war he would lose unless he came to terms with nature."[10]

Carson's prime target in *Silent Spring* was the chemical compound dichlorodiphenyltrichloroethane, aka DDT—a common compound used to rid neighborhoods all across America of pesky insects, particularly disease-carrying mosquitoes. DDT was also in widespread use throughout Africa and the Third World in an effort to rid those regions of deadly, mosquito-borne malaria. Carson portrayed the insecticide as lethal to both humans and animals—which is demonstrably false.

Despite its heavy doses of socialist induced junk-science, by the late sixties and early seventies, *Silent Spring* became required reading in high schools across America.

DDT AND ME

In 1983, I took my first job as a television meteorologist in the San Francisco market. One evening the station's anchorman told me of a captivating speaker he had just heard at a civic luncheon. The speaker was Dr. J. Gordon Edwards of San Jose State University, who began his talk by pouring a teaspoon of DDT into a glass of water and then drinking the glass empty. Edwards was a colorful entomologist who had taught at the university for decades. Besides being an expert in the insect world, he was also a famed mountaineer and a well-known conservationist who had made quite a name for himself by dramatically illustrating the lies of Carson's *Silent Spring*.

As a young man in 1962, Edwards was working with a team of researchers at Glacier National Park in Wyoming when he first read Carson's book, and was appalled by her allegations regarding DDT. Originally created by a German scientist in 1874, DDT was perfected as an effective, human-safe insecticide in 1939 by Swiss scientist Paul Muller (no relation to H. J. Muller), who won a Nobel Prize for his work. In many ways, DDT was a miraculous compound—inexpensive, nontoxic to humans, and extremely effective in eradicating targeted insects, while other forms of wildlife seemed to be immune to its effects.

Dr. Edwards had firsthand experience with DDT. In 1944, while serving as a U.S. combat medic stationed in Italy, his company was plagued with body lice. The fast-breeding insects were spreading typhus among the troops in some parts of Europe—frightening, given the fact that some thirty years prior, the disease had killed 3 million people in that very same part of the world. To curtail the developing epidemic, the chemists at Merck & Company in New Jersey produced the first five hundred pounds of American-made DDT and rushed it to the troops in Italy, a move that would seem unlikely if DDT had been—as Carson stated—a nerve agent.

Edwards's job was to dust the troops. "For two weeks I dusted the insecticide on soldiers and civilians, breathing the fog of white dust for several hours each day," he would later explain.[11]

As intended, the DDT applications worked, and the typhus outbreak was checked. It was estimated by the surgeon general that the DDT had saved the lives of at least five thousand soldiers. Confident and inspired by this experience, Edwards went on to earn his PhD in entomology from Ohio State University. His extensive research completely debunked the primary pillar of Carson's book.

Despite the flaws and fraud in *Silent Spring,* the United States Fish and Wildlife Service devotes numerous pages of its website to crown Carson as the queen of a movement, despite her association with well-known anti-American interests. According to the Fish and Wildlife Service, "Carson is credited with launching the contemporary environmental movement."[12]

POPULATION BOMBER: PAUL EHRLICH

A few years after Carson's book stole the minds of many, a second screed claimed more hostages in 1968—Paul Ehrlich's *Population Bomb*. Ehrlich, a professor at Stanford University, has authored many best-selling social engineering books during his decades at the ultraliberal San Francisco Bay Area institution, but the *Bomber* was his first hit. Like *Silent Spring*, it, too, became required reading in many public schools in the early seventies as Ehrlich falsely proclaimed, "The battle to feed humanity is over. In the 1970s and 1980s hundreds of millions of people will starve to death in spite of any crash programs embarked upon now."[13]

Ehrlich's antihuman message is a hybrid of radical thought originating with von Liebig's guano argument, Marx's *Laws of Matter*, Lenin's radical conservation program, and Robert Malthus's *Essays on Population*, written in 1798. Malthus, by the way, believed that unchecked population growth always exceeds the food supply. He contended that improving the lives of the lower classes or improving agricultural conditions were fruitless because these steps would only encourage humans to have more offspring, which would exacerbate the original problem. Malthus argued that as long as this tendency remains, the "perfectibility" of society will always be out of reach.

Ehrlich has long opined that the earth is being forced to support too many people who require too many resources and who produce too much pollution. For Ehrlich, the solution has never been promoting honest governments and freer economies in the Third World so that infrastructure could be constructed, clean water and sanitation made readily available, and successful farming practices implemented. Ehrlich's conviction has always boiled down to the problem of people—there are just too many of them—and his final solution has always been clear: "Population control is the only answer."[14]

Ehrlich's wild allegations include equating the earth's surplus of people with a *cancer* that needs to be eradicated: "A cancer is an uncontrolled multiplication of cells; the population explosion is an uncontrolled multiplication of people... We must shift our efforts from treatment of the symptoms to the cutting out of the cancer. The operation will demand many apparently brutal and heartless decisions."[15]

The method to Ehrlich's madness was revealed in his dictatorial action plan: "Our position requires that we take immediate action at home and promote effective action worldwide. We must have population control at home, hopefully through changes in our value system, but by compulsion if voluntary methods fail."[16]

While Carson convinced a generation that modern American liberty, ingenuity, free enterprise, and capitalism were the problems supposedly ruining the planet, Ehrlich introduced the solution: *a change in our value systems...by compulsion if other methods fail.*

While Carson's machinations were terminated by her early death in 1964, Ehrlich's assertions have continued to fester. In a September 2007 radio broadcast, Ehrlich pontificated, "Every scientist knows that we have too many people and that we should be reducing population size and that it's good for people—and women in particular—and for the environment, to have smaller families and to save lives."[17]

Marx and Engels couldn't have executed a better one-two punch.

CONVENIENT ECOLOGICAL DISASTERS

A series of ecological episodes insinuated that Carson and Ehrlich were messengers of enlightenment. The New York City garbage collectors' strike of 1968 was hailed as the greatest ecological disaster of the time and was hyped by eco-socialists as undeniable proof that mankind was trashing the planet. With only three television network newscasts to choose from at the time, news producers, who understood that alarming stories keep viewers fixed to the set, made sure that virtually every American witnessed trash piling up on New York's sidewalks, which they told us measured ten thousand tons per day.[18] It was "pollution on parade" for all to behold.

In January of the following year, a Union Oil drilling platform six miles off the coast of Santa Barbara, California, sprang a leak, allowing hundreds of thousands of gallons of crude oil to seep into the Pacific and wash ashore. Once again, the media provided Leftists with a convenient tool to demonize the U.S. lifestyle, as oil-coated birds were stuck in the same muck that was used to power America's cars.

The nation's first outspoken congressional environmentalist, Wisconsin senator Gaylord Nelson, immediately flew out to California to see the action for himself. He returned to Washington, angered at the oil industry, vowing to "do something to wake America up."[19]

Several months later, in June of 1969, another event was etched into the American psyche when the Cuyahoga River in Cleveland, Ohio, became a symbol of a planet in disrepair. The accepted story is that the entire river was consumed in flames and burned for hours, but the truth is, the fire burned for fewer than thirty minutes, and no conclusive evidence of its cause has ever been determined, though it is widely accepted that the combination of industrial waste and floating debris somehow ignited beneath a train trestle. The entire blaze was extinguished so quickly that nary a photographer had time to snap a photo and the local media instead had to settle for a fireboat hosing down the charred trestle. That didn't stop *Time* magazine, however, from running a dramatic cover shot of the burning Cuyahoga River in their following edition—from a much more serious fire that occurred on the river back in 1952.

The ecological news events of '68 and '69 coincided with hundreds of thousands of hippies and young Marxist revolutionaries taking to the streets of America to protest the Vietnam War specifically, and capitalism in general. These were the largest protests ever witnessed in the U.S., and the nightly news broadcasts brought the drama into every living room in the country. A new movement had been born: a movement fueled by new narcotics, new music, and a new desire to "get back to the land," as the hippies were fond of saying. Their defining moment occurred on a farm in Woodstock, New York, in August of '69. The Woodstock Music and Art Fair drew two hundred thousand young people, who camped in the mud, got high, and got laid. Yeah, some of the music was good, but overall the event served up a three-day diet of anti-American slop. The concert was released as a popular film, and the soundtrack was sold on record and tape. Tens of millions of young Americans were eventually empowered by the message of Woodstock and became part of the rebellion aimed at wrecking Mom and Dad's USA.

EARTH GETS HER DAY

Seizing the Woodstock moment, Senator Nelson met with Professor Ehrlich a month later to discuss, we are told, "overpopulation," but the conversation certainly entailed much more.[20] Following their meeting, Nelson decided to employ a popular tactic he had developed, known as the "teach-in." His efforts were aided by a young Leftist named Denis Hayes. Hayes was a former student body president from Stanford with an effective track record for organizing anti-war, anti-America protests. While pursuing a master's degree in public policy at Harvard, Hayes had heard about the teach-in concept and sought out Nelson to help him take his strategy of infiltrating the classroom nationwide.[21]

I recall the ridiculous teach-ins from *my* youth. Scrapping the assigned curriculum for the day, teachers would have us sit cross-legged on the floor and "rap" about how America was an imperialist nation and why communism really wasn't such a bad form of government—it just needed to be implemented properly.

Thankfully, I didn't buy it.

"My God," Nelson said, following his meeting with Ehrlich. "Why not a national teach-in on the environment?"[22]

Years later, Nelson elaborated, "I was satisfied that if we could tap into the environmental concerns of the general public and infuse the student anti-war energy into the environmental cause, we could generate a demonstration that would force this issue onto the political agenda."[23]

Soon, Senator Nelson formally announced there would be a "national environmental teach-in" sometime in the spring of 1970.[24] Ehrlich and Hayes would play pivotal roles—Ehrlich as an academic agitator and Hayes as the rebellious organizer. After careful consideration, a name and date for the event were chosen: Earth Day would be celebrated each April 22.

Whenever I speak in major urban settings, I often ask the audience how many are originally from the former Soviet Union. Almost always, hands shoot up. I then ask those who raised their hands, "What's significant about April 22?" They immediately reply in unison, "Vladimir Lenin's birth date!"

It's a birth date they wish they could forget.

April 22 is the date of Lenin's birth, and on April 22, 1970, the murderous communist would have been one hundred. Selecting that date to "celebrate" Earth Day was clearly no coincidence. In light of Lenin's devotion to nature—even at the expense of his own people—the communist vanguard provided a model for the government's role in regulation of the environment. In addition, Lenin also perceived government-run education as the most efficient means of indoctrination: "Give us the child for eight years and it will be a Bolshevik forever," is the infamous quote attributed to him.

As a prequel to the inaugural Earth Day, a thousand people gathered at Santa Barbara City College to mark the anniversary of the Santa Barbara oil leak. Speakers included Ehrlich, who reflected that the spill represented everything he had been warning, namely, that America's swelling population and subsequent development was responsible for pillaging natural resources and ruining the environment. Denis Hayes was also there, invigorating the crowd and letting them know that in just three months a new educational opportunity would hit American schools, get loads of press, and usher in revolutionary change.

Nelson and Ehrlich were already known as nontraditional crackpots, but young Hayes was that and more. In a *New York Times* article published the day after that first Earth Day, titled "Angry Coordinator of Earth

Day," young Hayes bragged that five years earlier he had fled overseas because he "had to get away from America." Hayes was so committed to his anticapitalist cause that he made sure his organization did not even produce any Earth Day bumper stickers; "You want to know why?" he asked the *Times.* Because "they go on automobiles."

NIXON'S GREEN BLUNDERS

President Richard Nixon took office in 1969 and inherited a cultural mess. As a well-known former adviser to Nixon once privately explained to me, Nixon "wanted to be liked," and thus attempted to pacify the hippies by extending an ecological olive branch. The appeasement began with a legislative act that was passed with little fanfare by Congress in 1969, and then was quietly signed into law by Nixon on the first day of 1970: the National Environmental Policy Act (NEPA). Essentially, the purpose of NEPA is to ensure that environmental factors are weighted equally when compared to other factors in the decision-making process undertaken by all federal agencies. As the preamble to NEPA states, the act was undertaken, "to declare national policy which will encourage productive and enjoyable harmony between man and his environment; to promote efforts which will prevent or eliminate damage to the environment and biosphere and stimulate the health and welfare of man; to enrich the understanding of the ecological systems and natural resources important to the Nation."

NEPA's muscle originates in its requirement of federal agencies to prepare an environmental impact statement (EIS) to accompany reports and recommendations for funding from Congress. Often, the agencies responsible for preparing an EIS do not compile the document directly, but outsource this work to private consulting firms with expertise in the proposed action and its anticipated effects on the environment. Thus, an EIS is only as reliable as the scientists providing the data, allowing an atrocious opportunity for highly biased findings based on junk science. Because the final EIS documents are often very technical and hundreds or thousands of pages in length, the likelihood that anyone in Congress would ever read a final report is virtually nil. The result is that now environmental impact statements are often the environmentalists' method of choice to kill any project they don't like.

According to an assessment of NEPA, written twenty-seven years later by the Clinton-Gore administration, "NEPA continues to provide a broad mandate for federal agencies to create and maintain conditions under which man and nature can exist in productive harmony and fulfill the social, economic and other requirements of present and future generations of Americans."[25]

Nixon's next gift to the radicals was his signing of the Clean Air Act of 1970. While this was not the first foray into the federal government becoming involved in pollution issues, it was the most overreaching and followed a one-two-three punch that countered the original intent of the Constitution. The Air Pollution Control Act of 1955 was the first federal legislation involving air pollution and provided funds for air pollution *research.* The Clean Air Act of 1963 established a program within the U.S. Public Health Service seeking air pollution *control.* In 1967, the Air Quality Act was enacted to expand federal government involvement via *enforcement.* In accordance with this law, enforcement proceedings were initiated to halt *interstate air pollution.* So, if it was deemed that air pollution from an industrial source, like a steel mill in Pittsburgh, Pennsylvania, was impacting the air quality of southwestern New York, the Pittsburgh mill would be in violation.

Nixon's Clean Air Act expanded the '67 law to autocratically allow the federal government to limit both industrial and mobile sources of air pollution (as in cars and trucks). Specifically, six distinct pollutants were targeted: ozone, sulfur dioxide, nitrogen dioxide, carbon monoxide, lead, and respirable particulate matter.

To enforce these new extreme laws, Nixon went a step further by playing the card of executive privilege. On July 9, 1970, Nixon authorized "Reorganization Plan Number 3." The order proclaimed: "Our national government today is not structured to make a coordinated attack on the pollutants which debase the air we breathe, the water we drink, and the land that grows our food. Indeed, the present governmental structure for dealing with environmental pollution often defies effective and concerted action."[26] Nixon's simple proclamation brought about a quick restructuring of the federal government and birthed the Environmental Protection Agency (EPA). Today the EPA is eighteen thousand full-time employees strong, with an annual budget of $10 billion. It has become the government tool of choice for regulating the air, the water, the auto industry, America's manufacturing base, lightbulbs, sewers, rodents, snakes, snails, minnows, and marshlands. The EPA regularly runs roughshod over states' rights and is the federal government's best excuse for confiscating private property. The EPA might as well stand for Endangering People's Activities. Creating the agency was Nixon's biggest mistake, and arguably one of the worst ideas ever forwarded by a U.S. president. It was just one of the many environmental hustles that took place in sixties and seventies.

From a legal perspective, what Nixon did in signing the Clean Air Act was flatly unconstitutional. Article 1, Section 8 of the Constitution limits the power of the federal government to the basics, none of which includes regulating the atmosphere. If a particular state wants to create air pollution laws, it may. Contrary to common perceptions, many measures of environmental quality were already improving before the advent of federal environmental laws and were delivering remarkable changes for the better. An honest look back at history reveals that Cincinnati and Chicago were the first cities to adopt effective smoke control ordinances as early as 1881. Following the Second World War, industrial cities began addressing their pollution problems, and in some cities, such as Pittsburgh—the "Iron City"—the business community played a leading role in supporting such regulation. Throughout the 1950s and 1960s, state and local governments began to recognize the importance of environmental quality and adopted first-generation environmental controls. Some states' efforts were more comprehensive and more successful than others, and different states had different priorities, but that is the beauty of states' rights—they can make those decisions for themselves. However, Nixon's Clean Air Act and founding of the EPA created a multiheaded, liberty-sapping monster that reinterpreted the Constitution. Suddenly the Article 1, Section 8 clause giving Congress the power "to regulate Commerce with foreign Nations, and among the several States, and with the Indian Tribes," was twisted to provide the feds with the legal muscle to control the air we breathe in—and ultimately even the air we breathe *out*, as in carbon dioxide.

CHIEF SPOOF-UM

Several months after the establishment of the EPA, with the environmental cause suddenly all the rage, an organization called Keep America Beautiful created a memorable public service announcement that aired on television stations across the country. In the seventies, the Federal Communications Commission required that TV stations air a certain number of PSAs promoting nonprofit organizations. Usually thirty to sixty seconds in length, most of these announcements were poorly produced and real audience turn-offs. Nonetheless, local station management had an obligation to run so many per day. Thus, when a good one came along, program directors would give the spot an abundance of airtime.

Such was the case with the "People Start Pollution, People Can Stop It" campaign. The initial airdate for the first ad was Earth Day 1971. The television commercial created to launch the campaign was especially well produced, extremely dramatic, and eventually honored by *Ad Age* magazine as one of the top advertising campaigns of the twentieth century. Given that this was before the proliferation of cable TV and that there were usually only five or six stations available to most viewers, everyone in the United States saw this PSA *many* times over—in fact, it ran on stations for the next ten years. The impressionable spot touched a nerve that brought a tinge of guilt to even the most calloused heart.

A stately, ponytailed American Indian dressed in traditional buckskin opens the minute-long ad, silently paddling his canoe up a serene river, accompanied by a dramatically rising orchestral score. In concert with the music, the harmful effects of humankind are incrementally revealed: trash floating past the canoe, towering factory smokestacks belching clouds of black in the distance, the riverbank strewn with junk. Climbing out of the canoe, the Indian deliberately strides up the riverbank, clearly troubled by the debris. His brisk walk leads to a highway embankment, where automobiles whiz by. The tanned-skin man appears to be an anachronism by the side of the highway. As one vehicle speeds by, an occupant throws a sack of trash out the window. It splatters on the ground like a rotten pumpkin all over the Indian's leather moccasins. He watches the vehicle race down the road as the camera cuts to a close-up of his sad, weathered face and a single tear rolling down his cheek. An unseen narrator says in a troubled tone, "People start pollution. People can stop it."[27]

America was smitten with this authentic, crying Native American. The general consensus was, "We *are* screwing up this planet!"

Soon we learned that this iconic figure's name was Iron Eyes Cody, a well-known Hollywood actor who did cameos and bit parts in a bevy of films and TV shows. Because of the Keep America Beautiful ad, Iron Eyes became the most famous Native American of the day.

However, in the late nineties, we discovered that the PSA, like most aspects of the environmental movement, was actually an outlandish fraud. Good Chief Iron Eyes was not a Native American after all, but an Italian from Louisiana, whose real name was Espero DeCorti. The trashed river was staged. Special effects were used to compose the pollution billowing from the smokestacks. Even the tear streaming from DeCorti's iron eye was nothing but Hollywood glycerin.[28] America had been punked.

As an aside, it somehow seems fitting that one of Iron Eye's—or Espero's—last acting roles was as an Indian on the ABC-TV show *Fantasy Island*.

SELLING THE WEATHER

America was falling for the Marxist pollution ploy hook, line, and sinker. However, for the eco-radicals, having public acceptance, media buy-in, and their own federal agency was hardly enough—their long-term goal was to commandeer the entire U.S. government and move it into a socialist paradigm. The screws would need to be progressively tightened. After all, what if all the pollution was cleaned up? How would they be able to demonize Life, Liberty, and the Pursuit of Happiness? A bigger crisis had to be concocted—one that really could prohibit development, inhibit property rights, convince people to have fewer children, and redistribute American wealth, domestically and internationally. Since the media was proving to be a willing coconspirator, an advanced lever needed to be fashioned and pulled.

How about the *weather*?

People have always been fascinated with the weather—and weather sells. Having been a TV meteorologist in both small markets and large—including nationally—I can tell you that the weather segment on the evening news is usually the most-watched portion of the broadcast. And big weather events always pack in the audience.

Believe it or not, the first person to use weather as a tool to trash humanity said this—in 1883:

> But inexorably the time will come when the declining warmth of the sun will no longer suffice to melt the ice thrusting itself forward from the poles; when the human race, crowding more and more about the equator, will finally no longer find even there enough heat for life; when gradually even the last trace of organic life will vanish; and the earth, an extinct frozen globe like the moon, will circle in deepest darkness and in an ever narrower orbit about the equally extinct sun, and at last fall into it.[29]

That lengthy sentence was penned by Frederick Engels. Some geologists at the time were advancing a theory that held that the earth was heading into another ice age, and Engels conveniently used their proposition to further his opposition to capitalism. No popular author would hit the subject again until 1968, when Paul Ehrlich took a shot at it in *Population Bomb*: "But even more important is the potential for changing the climate of the Earth. All of the junk we dump into the atmosphere, all of the dust, all of the carbon dioxide, have effects on the temperature balance of the Earth."[30]

And what types of weather changes were occurring at the time of Ehrlich's writing? Global *cooling*. The average temperature on planet Earth had been steadily decreasing since 1940.[31] Like Engels years before, the modern eco-scientists blamed the changes on pollution. Their flimsy theory went something like this: the air pollution created by mankind's cars and industry was suspended in the air, obscuring the sun's rays and causing solar radiation to be reflected back into space, causing the earth's temperature to cool.

And predictably, the media ran with it—for several years:

> The facts of the present climate change are such that most optimistic experts would assign near certainty to major crop failure in a decade... [If nothing is done] mass deaths by starvation and probably anarchy and violence" [will result].
>
> —*New York Times*, December 29, 1974

> As for the present cooling trend a number of leading climatologists have concluded that it is very bad news indeed.
>
> —*Fortune Magazine*, February 1974

> Climatological Cassandras are becoming increasingly apprehensive, for the weather aberrations they are studying may be the harbinger of another ice age
>
> —*Time*, June 24, 1974

> There are ominous signs that the earth's weather patterns have begun to change dramatically and that these changes may portend a drastic decline in food production—with serious political implications for just about every nation on Earth. The drop in food output could begin quite soon, perhaps only ten years from now.... Last April, in the most devastating outbreak of tornadoes ever recorded, 148 twisters killed more than 300 people and caused half a billion dollars damage in 13 U.S. states.
>
> —"The Cooling World," *Newsweek*, April 28, 1975

> Scientists Ponder Why World's Climate Is Changing; A Major Cooling Widely Considered to Be Inevitable
>
> —Headline, *New York Times*, May 21, 1975

> The threat of a new ice age must now stand alongside nuclear war as a likely source of wholesale death and misery for mankind.
>
> —Nigel Calder, editor, *New Scientist*, 1975

Certainly the tone of these headlines sounds familiar. Replace "cooling" with "warming," and these quotes from the seventies sound like the same frightening lines we are bombarded with today. In his best-selling 1976 book, *The Cooling*, Lowell Ponte went so far as to claim, "The cooling has already killed hundreds of thousands of people in poor nations."

Interestingly, Ponte's *Cooling* was endorsed by a young, up-and-coming atmospheric activist, Stephen Schneider, who wrote on the book's cover: "This well-written book points out in clear language that the climatic threat could be as awesome as any we might face, and that massive worldwide action to hedge against that threat deserve immediate consideration." Keep in mind, with great passion Schneider was advocating "massive worldwide action to hedge against" *global cooling*.

Thirteen years later, Schneider, who became a celebrated professor of environmental biology and global change at Stanford, would release his own much-heralded book, *Global Warming*. The year his book came out, Schneider puffed:

> We are not just scientists but human beings as well. And like most people we'd like to see the world a better place, which in this context translates into our working to reduce the risk of potentially disastrous climatic change. To do that we need to get some broad-based support, to capture the public's imagination. That entails loads of media coverage. So we have to offer up scary scenarios, make simplified dramatic statements, and make little mention of any doubts we might have.[32]

Dreaming up scary scenarios, making simplified dramatic statements, and lying through your teeth regarding any doubts you might have isn't science; it's pathological chicanery. Nonetheless, the stage was now set to take this hijacking of the atmosphere to the next level.

As an aside, Schneider and I were professionally acquainted. He was disappointed that I had become a "right wing denier," while I was sad he was a big-time socialist supporting an agenda to dismantle liberty. He died of a heart attack in 2010 at age sixty-five.

ENTER THE MASTER

Pass him on the street and you'd never give Maurice Strong a second glance. He is a short, overweight, gray-haired, soft-voiced man with a salt-and-pepper toothbrush mustache. Born in 1929, Strong grew up in a humble working-class family in Manitoba, Canada, and dropped out of school at the age of fourteen. He went on to become a multimillionaire and the United Nations' chief environmental operative.

Quite frankly, Strong's curriculum vitae is even more curious than Barack Obama's.

According to Strong's official website, as a teenager he worked on passenger ships, railroads, and with the native Inuits in the Arctic before somehow obtaining a job as a humble clerk in the newly formed United Nations in New York in 1947—he was barely eighteen.[33]

By all accounts, it was in New York where Strong met David Rockefeller, grandson of the billionaire founder of Standard Oil, John D. Rockefeller. David Rockefeller was a well-heeled radical. He graduated from Harvard University in 1936, doing his senior thesis on Fabian socialism. Fabian socialists believe that public opinion can be changed by properly educating an elitist class who can then lead government policy reforms. Fabians also hold dear the unionization of workers, a national minimum wage, rent control, progressive taxes, and the federalization of land (in other words, Fabian Socialists have much in common with today's American liberal politicians). In 1940, Rockefeller received his PhD from the University of Chicago (founded by his grandfather); his dissertation was titled "Unused Resources and Economic Waste."

In 1946, Rockefeller became the family's first banker when he joined the staff of the longtime family-associated Chase National Bank. The bank's chairman was his uncle Winthrop Aldrich, the son of the powerful U.S. senator Nelson W. Aldrich, and the brother of Rockefeller's mother. Chase National subsequently became the Chase Manhattan Bank in 1955, and is now known as JPMorgan Chase.

Conspiracy theorists have always found a target in David Rockefeller for supposed ties to secret international cabals set upon influencing the global economy and politics. And in 2002, their suspicions were confirmed with the publication of Rockefeller's autobiography, *Memoirs*. On page 405 Rockefeller states: "Some even believe we are part of a secret cabal working against the best interests of the United States, characterizing my family and me as 'internationalists' and of conspiring with others around the world to build a more integrated global political and economic structure—one world, if you will. If that's the charge, I stand guilty, and I am proud of it."

The records are unclear as to how long Strong worked at the UN, but certainly within less than a year he suddenly quit. Strong's website states that he "decided the best course...would be to return to Canada and try to develop there the qualifications that would enable him to return to the United Nations in a more substantive role."[34]

Strong went back home, but never did attend college. Instead, most suspect David Rockefeller assisted in opening the doors for Strong to begin one of the more phenomenal financial success stories in corporate history. Strong started in the oil business in the 1950s. By age thirty-five he was president of the Power Corporation of Canada. He went on to cofound Petro-Canada oil company, which at the time was nationally owned, and eventually, he became head of Ontario Hydro, North America's largest utility company (also state-owned).

Oddly, in 1970, Strong left his wildly successful Canadian business pursuits to return to the United Nations, where he began a quick succession of high-level appointments. The following year, he became a trustee for the Rockefeller Foundation. In 1972, he chaired the UN's first environmental gathering, the Stockholm Conference. Prior to the confab, Strong told the BBC, "Stockholm must usher in a new era of international cooperation."[35]

The Stockholm Conference gave birth to the United Nations' most aggressive program for assaulting America, UNEP—the United Nations Environment Programme. Strong was named its Executive Director.

Under Strong's leadership, UNEP began to brazenly unfurl its socialist agenda. In 1976, Strong described himself to Canada's *Macleans* magazine as "a socialist in ideology, a capitalist in methodology."[36] In that same interview, he warned that if the nations of the world would not heed his environmentalist warnings, Karl Marx's law of matter—particularly the law of negation—will force the earth to collapse into chaos. "Do we really want this? Do we want Marx to be proven right, after all?" Strong asked.[37]

The same year, the UN sponsored the World Conference on Human Settlements (known as Habitat I). Here Strong's UNEP boldly declared its Marxist view of private property and issued plans to eradicate it:

- "Private land ownership is a principal instrument of accumulating wealth and therefore contributes to social injustice. Public control of land use is therefore indispensable."[38]

- "Public ownership of land is justified in favor of the common good, rather than to protect the interest of the already privileged."[39]

- "Zoning and land-use planning [are to be used] as a basic instrument of land policy in general and of control of land-use changes in particular."[40]

- "Fiscal controls [are to be employed], e.g. property taxes, tax penalties and tax incentives [in order to eradicate private lands]."[41]

Maurice Strong would go on to build a well-crafted global coalition aimed at using a manufactured atmospheric crisis to bring down free markets and alter representative republics like the United States. Ehrlich would continue his antihuman diatribes, and in 1977, he would pen a missive with his protégé, John Holdren, titled *Ecoscience, Population, Resources, Environment*. In that work, the two proclaim, "Indeed it has been concluded that compulsory population-control laws, even laws requiring compulsory abortion, could be sustained under the existing Constitution if the population crisis became sufficiently severe to endanger the society."[42]

Ehrlich would retread old ground with the 1990 release of his *Population Explosion*. In that treatise, the fiery author blames virtually every human catastrophe, both real and imagined, on overpopulation and Christianity. And guess who wrote Ehrlich's dust jacket endorsement? None other than future vice president, Nobel laureate, and Oscar winner Al Gore: "If every candidate for office were to read and understand this book, we would all live in a more peaceful, sane, and secure world."

Ehrlich and Schneider would go on to become trusted advisers to Gore. Denis Hayes would continue as the international leader of Earth Day. John Holdren would get an office in President Obama's West Wing as the White House director of science and technology policy. And Rachel Carson's *Silent Spring* would continue to sell like tofu at a vegan diner—you can find it on a shelf in virtually every bookstore in America.

THREE

UNVEILING THE AGENDA

I HAVE ALWAYS ENVISIONED THE INNARDS of the United Nations' headquarters as resembling Chalmun's Cantina—the fictional bar from the *Star Wars* universe, located in the pirate city of Mos Eisley on the planet Tatooine. Chalmun's was the haunt for a variety of weird-looking creatures of various alien races. While some of the cantina's odd patrons appear content sipping fancy beverages, smoking elaborate pipes, and listening to the tootling sounds of Figrin D'an and the Modal Nodes, others are ripe to take the head off the first hooligan who gives them a wrong glance.

Goony characters in some ways similar to those Chalmun's patrons populate the halls of the United Nations. I'm convinced many UN ambassadors and bureaucrats would barely bat an eye if they were suddenly beamed into the faraway fictional cantina. After all, the United Nations is the only place in the universe where communists, socialists, Islamists, strongmen, thugs, and dictators are able to cast a vote equal to that of countries beholden to principles of freedom and liberty. The UN has always called this "justice"—I call it travesty.

Founded in 1945, the UN has unmistakably failed in its primary mission to rid the world "from the scourge of war." However, they have fared much better in their secondary pledge "to promote social progress."

On its own, social progress is one of those slogans that sounds typically benign: "Yeah, sure. Social progress. I'm all for it." But like most left-oriented expressions, there is a sinister translation.

Social progress is a concept that was hashed out by Karl Marx's mentor, Georg Hegel. Hegel believed human progress was not about the inalienable rights of Life, Liberty, and the Pursuit of Happiness; those rights, he was convinced, were ideas born of irrational feelings—mere products of the mind. Instead, social progress, according to Hegel, is a freedom that can be achieved only when granted *by the state.* "The state," he declared, "in and by itself is the ethical whole, the actualization of freedom."[1]

Hegel theorized that if carried out properly, this government-based freedom would progress society into a communal world void of oppression, discrimination, and poverty. He also supposed that social progress promised living without the moral constraints imposed by religion and notions of absolute truth.

While Hegel regarded himself as a philosopher, Marx's personal reflection was that of an engineer designing the master plan for global transformation. Referring to his mentor, Hegel, Marx said, "Philosophers have hitherto only interpreted the world in various ways; the point, however, *is to change it.*"[2]

So, it's important to remember, when you hear the term "social progress" (or even the abridged version, "progress") in the context of political discourse, it is immediately recognized by those aligned on the political left as code for instituting a heavy-handed form of government that limits rights, controls the economy, redistributes wealth, and determines morality.

This is what the UN stands for. Their unstated but well understood goal is to institute such a system in the United States of America. Their master plan for accomplishing this stratagem is the green agenda.

BREWING THE GREEN AGENDA

The formal origins of the agenda began with a 1967 plan proposed by Sweden to hold a first-of-a-kind international environmental conference. At the time, the secretary general of the UN was U Thant of Burma. Thant was an admitted socialist (though he claimed he did not like communism's "violent tactics"[3]) who recognized that utilization of the environment could be a strategic means for achieving social progress. In May 1969, Thant presented a fiery speech before the United Nations in which he predicted the earth had only ten years to avert environmental disaster.[4] The following month, he blamed the bulk of the coming planetary apocalypse on the United States.[5] Working fervently to rally other member states to his cause, by the next year, Thant had garnered significant support, and the UN voted overwhelmingly to hold the world's first environmental conference in Stockholm in 1972. Maurice Strong was tapped as the man to lead the confab, both as the chief organizer and planner of the formal agenda. Strong chose the theme "Only One Earth."

In June 1972, official representatives from 113 countries, hundreds of nongovernmental organizations (NGOs), and thousands of media outlets assembled in Sweden to change the world. From the get-go, the rhetoric was impetuous.

Mrs. Indira Gandhi (her husband was *the* Gandhi—a man who espoused a type of Fabian socialism[6]) opened the conference, stating, "It is clear, that the environmental crisis which is confronting the world, will profoundly alter the future of our planet."[7] Former UN secretary general (and former Nazi officer) Kurt Waldheim hit the podium as well, using his bullying speaking style to proclaim, "The iron rule remains: our world is one, inseparable, and interdependent. It is this world that is threatened by the impact of man's unplanned, selfish and ever-growing activities."

But aside from the ecological balderdash, the conference's Declaration, crafted under Strong's supervision, clearly stated what the conference was really all about, social progress:

> Industrialized countries should make efforts to reduce the gap between themselves and the developing countries... A point has been reached in history when we must shape our actions throughout the world with a more prudent care for their environmental consequences... To achieve this environmental goal will demand the acceptance of responsibility by citizens and communities and by enterprises and institutions at every level, all sharing equitably in common efforts.[8]

The Declaration was nothing less than socialism cleverly cloaked in an ecological wrapper—"reduce the gap" between the rich and poor; "shape our actions" by surrendering to nature; "demand the acceptance of responsibility" through heavy-handed government regulations and laws. It was a corrupted prescription written with the goal to one day infect America.

At a special symposium for NGOs, featured speaker Paul Ehrlich praised Maurice Strong for his efforts to bring concerns of the environment and social progress to the international stage. "People in general have begun to realize we are in the middle of an environmental crisis," Ehrlich stated in a press conference. "I think Maurice Strong has done an absolute miracle of even getting people to come here and putting together this sort of conference. At the same time, we have to be extremely critical of the point of view that simply turns the old economic crank faster and is going to somehow close the gap between the rich and the poor."[9]

Afterward, in cowing to the demands of the conference, President Nixon was struck with a sort of Stockholm syndrome, telling the American people, "The United States achieved practically all of its objectives at Stockholm."[10] Nixon was not only a cheerleader for this conference, but he had even made formal proposals in advance, including one in which the United States government would seed a pot of cash to be used in forming a global environmental fund that could then be distributed to poorer nations, supposedly to aid in local pollution and sanitation matters. The UN quickly approved his proposition, and shortly after, Nixon issued an official statement saying, "The conference approved forming a $100 million United Nations environmental fund which I personally proposed last February."[11]

Nixon was able to twist enough weak arms in Congress to create a law, which he signed in 1973, handing over $40 million to the UN Environmental Fund over a five-year period.[12]

Stockholm was a seminal moment for the green agenda, but there was much more work to be done to further define green terms and carefully lay out a cardinal plan for domination of the U.S.

SPECIES FIRST, PEOPLE SECOND

For the next ten years, Maurice Strong would continue refining the green agenda. His ensuing entrée was presented in 1982 in a UN document known as the "World Charter for Nature." Strong and selected cohorts crafted this pact, succinctly stating that, in the tradition of materialism, mankind's place in nature was no more significant than any other species': "Every form of life is unique, warranting respect regardless of its worth to man, and, to accord other organisms such recognition, man must be guided by a moral code of action."[13]

Note the shrewd idea of turning environmental concerns into a moral issue. I will expand on this in great detail in chapter 6.

Next, the Charter attacked free markets and quality-of-life choices, stating, "In formulating long-term plans for economic development, population growth and the improvement of standards of living, due account shall be taken of the long-term capacity of natural systems."[14]

Essentially the World Charter presented the planet with an updated, hybrid version of Marx's laws of matter combined with Lenin's Decree on Land. All nonhuman species were now elevated to a unique status by the UN. It would only be a matter of time before this impaired mentality would successfully seep into the American consciousness.

Certainly you've heard of the many stories where proposed construction for a house, hospital, road, shopping center, or airport has been halted because an environmentalist claimed an obscure species would be threatened by the project. I recall an absolutely ridiculous case that occurred in San Francisco. In 2000, a dead garter snake was found in the construction zone for an extension of a commuter train line near the San Francisco airport. Construction of the project was immediately halted. From the get-go, the garter snake was an issue in this colossal public works project. In my opinion, if this had been a construction project for a job-creating, major retail center, for example, the plan never would have been approved. However, this commuter line is part of a long-term, sweeping green agenda mass-transit plan for which representative Nancy Pelosi—a green agenda priestess—was able to secure $30 million in federal funding in 1997.[15] So, naturally, in keeping with the spirit of the UN, these "long-term plans for economic development, population growth and the improvement of standards of living" were approved, and "regardless of its worth to man," the San Francisco garter snake would be protected, and a "moral code of action" would be employed to ensure its survival.

As a result, ridiculously expensive accommodations were made for the snake, all picked from the taxpayer's pockets. First, the California Department of Fish and Game shut the construction project down for eighteen days, during which time the unionized construction workers continued to get paid, costing the taxpayer $1.7 million.[16]

Then, a costly snakeproof fence was installed about the perimeter of the construction zone. Next, reptile experts were brought in who found seventy-five snakes that were "relocated" to habitat deemed safer. Professional "biological monitors" were also hired to remain on-site at all times during construction. All told, more than $6 million was spent to comply with the protestors' demands.

I'm not sure how easy it is for snake experts and biological monitors to get work, but this was a clear case of taxpayer dollars being used to provide temporary employment for green do-gooders with few marketable skills.

And the absurdities continued.

Prior to construction restarting, workers underwent a special "snake training" seminar, teaching them how to look for and recognize the San Francisco garter. In addition, the speed limit on the construction site was reduced to 10 mph (complete with "traffic monitors" armed with radar guns); further rules required that workers be driven into the construction zone by bus instead of driving their own cars.

After all this, two years later, another dead garter snake was discovered. Again, state wildlife officials demanded that work stop, and the dead snake was turned over to the snake monitor, who attempted to determine if the snake had died of natural causes or was killed by a construction worker. To the rational mind, all of this fuss and expense over a very common snake is lunacy. However, in the compendium of the green agenda, it's in keeping with two additional structural components: "sustainable development" and "social equity."

SUSTAINABLE DEVELOPMENT

Sustainable development is another concept developed by Maurice Strong. It was formally introduced in 1983 by way of a UN resolution calling for "the establishment of a special commission that should make available a report on environment and the global *problématique* to the year 2000 and beyond, including proposed strategies for sustainable development."[17] Strong was the point man on this elite committee tasked with establishing the priorities for the next millennium. The group was officially known as the World Commission on Environment and Development (often referred to as the Brundtland Commission, after its quasi-chairman, Harlem Brundtland of Norway). Its members labored quietly for four years before unveiling their lengthy manifesto, *Our Common Future*. The document details how establishing sustainable development will be the key to instituting global socialism.

Sustainable development is what drives the environmental movement. It's best described by environmentalists with analogies such as these:

- The use of natural gas as a source of electricity is unsustainable because it will eventually run out, plus it emits carbon dioxide, which is forcing climate change. However, the sun's rays are ever abundant and free; therefore the development of solar power is sustainable.

- Despite the fact that dams can be used to produce emission-free energy, and reservoirs provide precious stores of water, they are unsustainable because they harm a variety of species. The sustainable alternatives are low-flow showerheads and toilets, water rationing, and preventing farmers from growing water-intensive crops.

- Automobiles are an unsustainable transportation choice because they rely on carbon-emitting gasoline, a limited fossil fuel resource. While a bicycle is the best sustainable transportation option, if one must use motorized travel, mass transit is the most sustainable conveyance.

Tied at the hip to sustainable development is "social equity." The term is credited to H. George Fredrickson from the University of Kansas, who began using the idiom in 1968. Fredrickson describes social equity as a pillar of public bureaucracy.[18] According to Fredrickson, "Social equity emphasizes responsibility in government services... Social equity emphasizes responsiveness to the needs of citizens"[19]

To the left, social equity is the expected outcome of sustainable development. Social equity demands the guaranteed use of public funds to support the poor, needy, disadvantaged, disenfranchised, minorities, and lesser-skilled laborers. Social equity requires government bureaucracies to adapt to meet the needs of those in such categories. Social equity insists that all species are treated equally. It also seeks a one-class society (with special privileges for the rule-makers, of course). Social equity *is* socialism.

Illustrating my point, the actual words of *Our Common Future*, declare:

> Sustainable development is development that meets the needs... of the world's poor, to which overriding priority should be given.[20]
>
> [Sustainable] development involves a progressive transformation of economy and society... Even the narrow notion of physical sustainability implies a concern for social equity.[21]
>
> Sustainable global development requires that those who are more affluent adopt lifestyles with the planet's ecological means—in their use of energy for example.[22]

Here's how all of this works in the real world: environmental concerns are addressed through the lens of sustainable development, with a successful outcome being one that furthers social equity. For example, one of my best friends is in construction, and his company won a bid to do a major project for the City of San Francisco. His firm was to provide a city-owned building with an energy-efficient weatherproofing product that would reduce monthly utility bills.

My buddy showed up for the initial project meeting with another partner from his firm. Their company had done work in major cities across the country, but never for San Francisco, and he was in for a lesson on social equity on steroids. Thirty-five bureaucrats packed the room for the kickoff meeting, all armed with laptops. This, he told me, was unusual right off the bat, as a meeting such as this usually only requires perhaps a dozen folks. All city staffers were minorities, and the two women in charge of enforcing the city's Equal Opportunity statutes seemed especially hostile to the two Caucasian contractors.

My pal had no problem with the racial and ethnic differences present. San Francisco is what is referred to as a "minorities in the majority" city, so he was expecting to be one of the few white people in the room.

In keeping with the principles of social equity, all firms bidding on this job had to employ union labor, with paid hourly wages in line with what San Francisco determined "equitable." The bidding process made certain that the company getting the job would be working under a tight budget, with a slim profit margin.

And there was more, as the E.O. ladies laid down the law to my friend: "San Francisco has a minority-majority population; that means 55 percent of your on-site workers must be nonwhite," the bureaucrat stated.

"That won't present a problem," my buddy said. "Almost all our workers are Latino."

"I didn't say 55 percent Latino," the woman shot back. "Only 14 percent need to be Latino. The rest of your workers need to match the ethnicity structure of our city—and you'll need to hire a certain percentage of women too."

My buddy met Ms. Social Equity straight on.

"And where will we find these workers?"

"City job pool. We'll supply them for you."

He was also told that hiring a worker who is a gay minority would be counted twice. I'm not kidding.

Eventually, he was able to retain most of his staff because many of the recruits from the job pool flunked the mandatory drug test. Those who did pass required extra training and hand-holding, which cost his firm money. It will be his last job with San Francisco.

Nonetheless, his company delivered a sustainable product, and the installation of that product spawned social equity. My friend's company was forced to employ workers who otherwise may not have had any work for a while.

Again, sustainable development is the business plan for the green agenda, and social equity is the anticipated result. In fact, that's pretty much how Maurice Strong described sustainable development to the *Canadian Business Review* in a 1990 interview, explaining, "It's like putting our planet, Earth Incorporated, if you will, on a sound business basis."

Revealing the breadth of his plan, Strong continued, "Now, sustainable development is a broad concept and it must be defined in virtually every sector of human activity... Development requires economic growth but it requires the kind of qualitative economic growth that permits us to meet social and human needs, which of course are the purposes of growth in the first place."[23]

EARTH SUMMIT

Now that Maurice Strong's concept of sustainable development had been successfully pitched, his next assignment was to create a monumental event in which tangible marching orders for the simultaneous implementation of social progress, social equity, and sustainable development could be presented to the governments of the world. That event was the 1992 United Nations' Earth Summit.

Held in Rio de Janeiro, the Summit welcomed 108 heads of state, accompanied by some 10,000 members of the press. Thousands of NGO representatives were also in tow. President George H. W. Bush was there to represent the United States. Senator Al Gore was also a part of the U.S. delegation. Since 1990, his profile had been rising steadily among the green radicals since the *New York Times* published his open letter, "To Skeptics on Global Warming." Now he was being treated like a rock star as his book, *Earth in the Balance*, had just been strategically released in conjunction with the UN gathering.

During his opening speech to the gathering, Strong stated that gone were the days of national sovereignty; instead the world community needed to embrace a system of wealth transfer in order to ensure "environmental security":

> The world community must move towards a more objective and consistent system of effecting resource transfers similar to that used to redress imbalances and ensure equity within national societies. Financing the transition to sustainable development should not be seen merely in terms of extra costs, but rather as an indispensable investment in global environmental security.

Sounding like Marx attacking the bourgeoisie, Strong went on to rip the lifestyles of those living in industrialized nations:

> The wasteful and destructive lifestyles of the rich cannot be maintained at the cost of the lives and livelihoods of the poor, and of nature... This Conference must establish the foundations for effecting the transition to sustainable development. This can only be done through fundamental changes in our economic life and in international economic relations, particularly as between industrialized and developing

> countries. Environment must be integrated into every aspect of our economic policy and decision-making as well as the culture and value systems which motivate economic behaviour.

Here Strong presented a critical component to his concept of sustainable development, essentially stating, *if people want to live an opulent lifestyle they may, but they will have to pay*. Such payments would come in the form of higher costs, fees, taxes, and penalties.

The prescribed key to ridding the planet of the "destructive lifestyles of the rich" would be the mass acceptance of the Rio Earth Summit's signature contract, which would formally usher in the green agenda for the twenty-first century: Agenda 21.

AGENDA 21

Mention Agenda 21 to some and they imagine hearing the theme to *The Twilight Zone*. They don't believe it's real. However, as you'll see, Agenda 21 is not only real, but its goals are being progressively implemented in the United States. Want to know why former House Speaker Nancy Pelosi was so ecstatic about the passage of the health care bill? It's because universal health care is deemed *a right* according to Agenda 21.[24] It is also the reason Pelosi was thrilled with the passage of cap-and-trade legislation in the House of Representatives in 2009, and why she was miffed when it never got an up or down vote in the Senate. Agenda 21 implores each nation "to control atmospheric emissions of greenhouse and other gases and substances."[25] In fact, simply visit the Environmental Protections Agency's website and you'll find a quote from a former administrator of the agency speaking of Agenda 21's voluminous plan in glowing terms:

> This was perhaps the most remarkable achievement of the conference: an ambitious, 900-page action plan for protecting the atmosphere, oceans, and other global resources.
>
> —William Reilly, EPA Administrator, 1989–1992[26]

Rio's Agenda 21 presented the nations of the world with a template to employ sustainable development in all areas of life, and the United States signed on to the agreement. The document begins by stating that virtually all that ails the world can be cured through sustainable development:

> Humanity stands at a defining moment in history. We are confronted with a perpetuation of disparities between and within nations, a worsening of poverty, hunger, ill health and illiteracy, and the continuing deterioration of the ecosystems on which we depend for our well-being. However, integration of environment and development concerns and greater attention to them will lead to the fulfillment of basic needs, improved living standards for all, better protected and managed ecosystems and a safer, more prosperous future. No nation can achieve this on its own; but together we can—in a global partnership for sustainable development.

To illustrate how this one-world agenda has wormed its way into contemporary conversation, allow me to provide you with some poignant illustrations:

REDEFINING WEALTH

Agenda 21 attempts to redefine wealth, stating, "Consideration should also be given to the present concepts of economic growth and the need for new concepts of wealth and prosperity."[27] In other words, we need less development, fewer conveniences, and less affluence. This redefinition was best articulated by then senator Barack Obama, who infamously told Joe Wurzelbacher—Joe the Plumber from Ohio—"It's not that I want to punish your success, I just want to make sure that everybody who is behind you—that they've got a chance at success too... I do believe for folks like me, who have worked hard, but frankly, also been lucky, I don't mind paying just a little bit more than the waitress that I just met over there [who]... can barely make the rent, because my attitude is that if the economy's good for folks from the bottom up, it's gonna be good for everybody... I think when you spread the wealth around it's good for everybody."

USING GOVERNMENT TO CHANGE CONSUMPTION PATTERNS

Agenda 21 states, "Achieving the goals of environmental quality and sustainable development will require...changes in consumption patterns."[28] Indeed, "Governments themselves [can] also play a role in [determining] consumption...and can have a considerable influence on both corporate decisions and public perceptions."[29]

Since Americans are the biggest consumers on the planet, participants at Rio knew this portion of the Agenda was aimed squarely at the U.S. And they also knew that the most effective way to impact the American lifestyle was through taxation, surcharges, and mandates.

Hence, a few years later, when gasoline was selling at just over a dollar per gallon, Senator John Kerry proposed increasing the national gas tax by fifty cents per gallon. His motivation was to coerce the consumer into purchasing less fuel to reduce emissions he felt were causing global warming.

This is one of the reasons why congressional liberals push for carbon dioxide regulations. By taxing such emissions, the prices of gas, oil, coal, and fuel will artificially skyrocket. In response, the consumer will use less. This will result in a change of consumption habits, forcing people to purchase compact cars, smaller homes, and perhaps eventually have fewer children.

Even the food we choose to consume is under attack in the name of sustainable development. U.S. citizens like to eat beef, and cows require a lot of water and land. In a televised interview, Obama's Environmental Protection administrator, Lisa Jackson, admitted how the feds are attempting to influence the beef industry in order to control meat production and consumption. Jackson said that raising cattle for beef "causes more global warming than all the planes and automobiles in the world." She went on to say, "There are opportunities for huge greenhouse gas emissions in meat control, and you know, for a long time, Secretary Vilsac at [the Department of] Agriculture has been saying to the farming industry, 'Listen, there are opportunities here to continue to have food grown in this country...[and] to do it in a sustainable way.'"[30]

POPULATION CONTROL

According to Agenda 21, "Population policy should also recognize the role played by human beings in environmental and development concerns."[31] Additionally the document reveals that the population control method of choice is abortion. Cloaking their message with terms such as "curative health facilities" (a deceptive alias for abortion clinics), the Agenda declares, "Governments should take active steps to implement programs to establish and strengthen preventive and curative health facilities that include women-centered, women-managed, safe and effective reproductive health care and affordable, accessible services, as appropriate, for the responsible planning of family size."[32]

Secretary of State Hillary Clinton has always been a supporter of the United Nations and an avid proponent of abortion as a right. During remarks made at the fifteenth anniversary of the UN's International Conference on Population and Development, Mrs. Clinton announced "the launch of a new program that will be the centerpiece of our foreign policy." The new U.S. program was a $63 billion expenditure that will employ abortion as a means of "preventing millions of unintended pregnancies" in the Third World. "In the Obama Administration," Clinton said, "we understand there is a direct line between a woman's reproductive health and her ability to lead a productive, fulfilling life."[33]

In other words, according to the State Department, the new "centerpiece" of American foreign policy is all about funding the ghoulish underbelly of the Agenda.

GREEN JOBS

"Green jobs" was a concept initially introduced at Rio as "green works" and, as originally stated, was meant to create jobs for the poor, needy, disadvantaged, disenfranchised, minorities, and lesser-skilled laborers.

Agenda 21 declares, "'Green works' programs should be activated to create self-sustaining human development activities and both formal and informal employment opportunities for low-income urban residents."[34]

President Obama made this agenda item a reality almost immediately after taking office through billions of dollars in stimulus money doled out to various agencies. For example, in May 2009, housing secretary Shaun Donovan used some $4 billion to create green jobs for people who live in low-income housing units. In announcing the plan, Donovan said the green workers would be replacing windows, insulation, appliances, and lightbulbs.[35] Coinciding with Donovan's announcement, Labor Secretary Hilda Solis said that $500 million would be spent to train these workers.[36]

Additionally, buried in the 2009 America Clean Energy and Security Act, which passed Nancy Pelosi's House but failed in the Senate, was a federally mandated energy-efficient building regulations program that would have superseded *all* local and state codes. The plan would have been enforced by what I labeled in my book *Climategate* "the national, green goon squad," funded in part by revenues from future energy taxes, as well as by an annual $25 million from the Department of Energy "to provide necessary enforcement of a national energy efficiency building code."[37]

The legislation would have also authorized the secretary of energy to "enhance compliance by conducting training and education of builders and other professionals in the jurisdiction concerning the national energy efficiency building code."[38]

ALTERNATIVE ENERGY

In conjunction with the green works program, Agenda 21 demands that alternative, "national energy programs" be implemented "in order to achieve widespread use of energy-saving and renewable energy technologies, particularly the use of solar, wind, biomass and hydro sources."[39]

Of course, we see this push occurring all around us in America, and later we will discuss alternative energy at great length. But suffice it to say, the real push for renewables is to prevent anyone from selling natural resources, like fossil fuels, for profit. Social equity is achieved through the creation of jobs stemming from the manufacture and installation of the solar panels and wind towers. This point is best made by Van Jones, Obama's short-time "Green Jobs Czar."

In a 2008 radio interview, Jones precisely articulated the role of alternative energy in transforming the American economy. "We now have to move in a very different direction," he said. "And key to that will be basing the U.S. economy...on clean energy and the clean energy revolution that would put literally millions of people to work, putting up solar panels all across the United States, weatherizing buildings so they don't leak so much energy and put up so much carbon, building wind farms and wave farms, manufacturing wind turbines...to help save the world."[40]

In another interview, Jones further revealed the plan: "We are saying we want to move from suicidal gray capitalism to some kind of eco-capitalism...so the green economy will start off as a small subset and we're going to push it and push it and push it through until it becomes the engine for transforming the whole society."[41]

MASS TRANSIT

In 2009, the Obama administration announced plans to construct ten high-speed rail lines covering virtually every part of the United States. The administration requested $53 billion for the massive project, with the various states picking up the tab for near-equal amounts. Never mind that we already have a failed national railroad system known as Amtrak. At last count, forty-one of the forty-four Amtrak routes were losing money annually, with some lines losing up to $462 per passenger.[42]

Agenda 21 states that countries shall "adopt urban-transport programs favoring high-occupancy public transport in countries,"[43] and make sure these major government-funded construction projects "promote the use of labor-intensive construction and maintenance technologies which generate employment in the construction sector for the underemployed labor force found in most large cities."[44]

And there's the huge social equity upside to building these massive railways: jobs.

BICYCLE AND JOGGING PATHS

Certainly you've noticed the countless recreational walking, riding, and jogging paths that have popped up seemingly everywhere. Conducting a search on the federal government website that tracks the use of stimulus funds, there were 13,804 bicycle, jogging, and pedestrian paths funded by the Stimulus Bill, formally known as the American Recovery and

Reinvestment Act of 2009, totaling billions of dollars.[45] I regularly see some of these well-constructed paved paths here in California and note that they boast very few, if *any*, users. However, they sure did allow for plenty of temporary "green jobs."

According to the Agenda, governments shall "encourage non-motorized modes of transport by providing safe cycleways and footways in urban and suburban centers."[46] Again, more "green" jobs.

PREVENTING DEVELOPMENT

Preventing land from private development is the kingpin item of Agenda 21. As the Agenda states, governments shall "formulate appropriate land-use policies and introduce planning regulations specially aimed at the protection of eco-sensitive zones against physical disruption by construction and construction-related activities."[47] This plank from Rio has done more to abolish physical private property in the United States than any other governmental policy.

And this is where, seated at a prominent table in Chalmun's, we meet Al Gore.

FOUR

GORE'S WELT ON AMERICA

I HAD JUST CONCLUDED my local radio broadcast in San Francisco when my cell phone began to buzz. The caller ID revealed it was an acquaintance from a charity with which I was involved.

"Hey, Dave," I greeted the caller. "What's goin' on?"

"Brian, I was just listening to your show and you were talking about Al Gore."

Indeed, I had been talking about the former vice president. I was telling my audience of his numerous connections in the Silicon Valley and how he was situated to make millions off of the global warming scare. Dave had called to set me straight on some of my facts.

To be honest, though I knew Dave (not his real name) within the context of the charity, I really didn't know what he did for a living. I knew he was a former special-ops guy in the military, and I also figured he had earned a little money post-military, but beyond that I had never asked him for specifics.

However, I was stunned when in his next sentence he informed me he had done some work in the Silicon Valley with Al Gore. You see, Dave's day job is that of a hedge fund manager, and one of his investment groups included a member named Al Gore.

"You're kidding me—you've worked with Gore?"

"Yes, and you've got it wrong about him. For starters, Gore stands to make *billions,* not millions—which he's already raking in—off of global warming."

Dave instructed me to meet him at a Starbucks near Apple's headquarters in Cupertino the next day. Once there, Dave set before me a stack of open-source documents illustrating how Gore was set to make billions off of green technology and cap-and-trade. "You've been working so hard to connect the dots. I'm saving you a year's worth of research," Dave said.

Our meeting occurred in 2009, and much of Dave's data made it into *Climategate.* Ever since losing the presidency in 2000, Gore had successfully positioned himself to profit off of the climate scare.

As we talked, Dave told me the details of a boardroom tussle he'd had with Gore regarding global warming. "I've gone back and forth with him, Brian. The guy's piggish—he's not the dashing young senator he used to be. He's unkempt, a bully, and doesn't like to be found wrong."

"Doesn't surprise me. Hard-core socialists never admit they're wrong."

I could tell Dave wanted to agree with me, but was without evidence; so he held his tongue.

"You do know about his ties to the former Soviet Union, don't you?" I asked.

Dave's eyes widened as I shared the story.

THE COMMUNIST IN GORE'S CLOSET

Albert Gore Jr. is the son of Albert Sr., an influential Tennessee U.S. representative and senator who served in Congress from 1939 to 1971.

One of Al Sr.'s close friends was a man named Armand Hammer. Hammer's background was more than curious. His father, Julius Hammer, was born in Russia in 1874. At sixteen, Julius moved to New York with his family. He went on to attend medical school at Columbia College, became a doctor, and later gained a seedy reputation as an abortionist. In addition, he was an aggressive businessman who ran eight drugstores with proceeds reportedly used to support the Bolshevik Party back in Mother Russia—a party that included the likes of Vladimir Lenin, Leon Trotsky, and Joseph Stalin. Not surprisingly, Julius became well known as one of the early influential leaders of the Communist Party in America.

In 1898, Armand was born into this nest of anti-American values, in

Manhattan. In time, he would continue the relationship his father had established with communist Russia.

In 1996, author Edward Jay Epstein wrote the most thorough work I have read on Armand Hammer, titled *Dossier: The Secret History of Armand Hammer.* Drawing from FBI and Russian intelligence documents, Epstein illustrated Hammer's extensive business dealings with Russia and the former Soviet Union, dating back to Stalin's murderous reign. Armand Hammer apparently laundered millions of dollars for the Soviet Union in bogus business transactions. He also partnered with Russia in mining operations, the manufacturing of heavy equipment, and even the production of pencils for Stalinist Russia. He traded fur and trafficked in Czarist art, apparently both real and phony. At one point, in a bizarre exchange of communist conversation, Vladimir Lenin told Stalin that Hammer was a "path leading to the American business world, and this path should be made use of in every way."[1]

Confirming his valuable service to the Soviet Union, Armand Hammer received Russia's highest honor, the Order of Lenin award. Though the accolade was usually presented to Soviet citizens for "outstanding services rendered to the State," there were several foreign recipients, including Cuban dictator Fidel Castro and ghoulish Yugoslavian strongman Josip Broz Tito. Other provocative members of the Order include the inventor of the AK-47, Mikhail Kalashnikov, and notorious Soviet spy Vilyam Genrikhovich Fisher, who was found guilty of running a long-term espionage operation out of New York in 1957.

That same year, Hammer made his biggest bucks by becoming CEO of U.S.-based Occidental Petroleum. Over the next decade Hammer's reputation as a shrewd dealmaker paid off as he was able to open doors for Occidental to expand internationally, including gaining exploration rights in Libya in 1965. In 1979, the United States declared Libya a state sponsor of terrorism, and by 1986, severe economic sanctions imposed upon Libya by the U.S. forced Hammer to suspend operations there.

Enter the relationship between Armand Hammer and Albert Gore Sr. Initial details of their meeting remain sketchy, but about the time that Gore Sr. moved from the House of Representatives to the Senate, in 1952, Hammer made him a partner in an Angus cattle–breeding partnership, which, according to Epstein, netted Gore "a substantial profit."

Former Tennessee governor Ned McWherter, a Gore family friend, was skeptical of Albert Sr.'s success in a difficult business and once said, "I've sold some Angus in my time too, but I never got the kind of prices for my cattle that the Gores got for theirs."[2]

As senator, Albert Gore Sr. returned the financial favor and opened many doors for Hammer. He was able to convince the Commerce Department to sponsor Hammer on a trip to Moscow. Gore also tried, but failed, to see Hammer appointed as an emissary to Berlin. The senator also arranged for Hammer to negotiate a few matters for the State Department, the Defense Department, and the FBI. Gore Sr. even attempted to help Hammer secure a long-term lease on a chemical manufacturing plant in West Virginia, owned by the U.S. Army. Hammer stated that his desire for the plant was to make it a general base of operations for his many ventures, including the exportation of synthetic nitrogen and ammonia-based fertilizer (the kind that can be used as a key ingredient in the making of bombs). The deal was stopped by the House Armed Services Committee. Payback for all Gore Sr.'s assistance to Hammer occurred soon after the senator failed to win reelection in 1972. Hammer offered Gore a sweet opportunity to buy into a below-market land sale in rural Carthage, Tennessee, that included mineral rights to a zinc mine on the property. Gore accepted the deal.

Gore Sr. would be further remunerated with a directorship on the board of an Occidental subsidiary, Island Creek Coal Company—a position reportedly worth five hundred thousand dollars annually.[3]

Interestingly, shortly after Hammer sold the zinc field to Gore Sr., a deed was fashioned giving the property, mineral rights, and twenty thousand dollars in annual royalties to Al Jr.[4] This came to light when Al Gore, by then a champion of the environment, was running for president in 2000.

Things turned sour for Hammer in 1975, when he pleaded guilty to three counts of making illegal campaign contributions—fifty-four thousand dollars in hundred-dollar bills—to a Nixon fund-raiser. Apparently his hope in supporting Nixon was that Tricky Dick would be the man to normalize relations with the Soviet Union, a move that would greatly benefit Hammer financially. President Ronald Reagan—no friend of communists or those sympathetic to their cause—refused to pardon Hammer. However, he was pardoned by President George H. W. Bush in 1989.

ENTER ALBERT JR.

After years of trying to find his place in life, including subpar work as a Harvard undergrad, followed by a brief stint as an Army journalist (he received an early release), a miserable time at Vanderbilt University's School of Religion (where he flunked out), and finally an attempt to become a lawyer at Vanderbilt's School of Law (he didn't finish), Al Gore Jr. decided to take advantage of his name recognition and run for Congress in Tennessee. He was elected as a pro-life Democrat in 1976, and by 1984, was elected as a pro-choice senator.

As Al Jr. began to navigate up the political ranks, his dad's old chum Armand Hammer began to provide favors. The Hammer family and its corporations made maximum legal donations in all of Gore's campaigns, according to Mr. Hammer's former personal assistant, Neil Lyndon.[5] Young Al was often seen dining with Hammer and Occidental lobbyists. According to Lyndon, "Separately and together, the Gores sometimes used Hammer's luxurious private Boeing 727 for journeys and jaunts." The former Hammer aide noted that the "profound and prolonged involvement between Hammer and Gore has never been revealed or investigated."

In 1987, Armand Hammer received a humanitarian award in Moscow from International Physicians for the Prevention of Nuclear War. The irony was bizarre. Russia was still ruled by a tyrannical regime beholden to the tenets of communism, and possessed a nuclear arsenal large enough to destroy much of the planet. Not only were they hosting an international conference to protest the use of nuclear weapons, but they were presenting a humanitarian award to a U.S. citizen whose president had recently declared the Soviet Union an "evil empire" and earlier that year challenged their leader, Mr. Gorbachev, to "tear down this wall."

But the paradox doesn't stop there. America's newest rock-star senator, Al Junior, accompanied Hammer to the conference and even delivered a speech to the group. Gore's message was well received by his communist hosts: not only should nuclear arsenals be cut, he declared, but so must conventional arms. Some covering the event were stunned by the "political risk" Gore took in speaking at the conference.[6]

Eventually, Gore's Moscow trip would turn out to be a résumé-enhancer. After being sworn in as Bill Clinton's vice president in January

1993, Gore was made the point man for Russian relations. And because of his standing as a green giant, Gore was immediately cut loose to act as a quasi-ambassador *from* the United Nations, to implement the goals set forth in Agenda 21, particularly those involving sustainable development.

GORE IMPLEMENTS AGENDA 21

In keeping with Agenda 21's promise that "each country should develop integrated strategies to maximize compliance with its laws and regulations relating to sustainable development,"[7] Gore used his considerable cache to launch a major three-day conference on the topic that spring in Louisville, Kentucky, hosted by the Blue Grass state's governor, Brereton Jones.

To promote the event, President Clinton assembled a conference call with Democratic governors and mayors. Clinton succinctly summarized a key facet of sustainable development in layman's terms, stressing to Governor Jones "[an important] point I want to make about your conference coming up in May, on sustainable development. One of our great challenges is to try to figure out how to improve the environment and improve the economy at the same time... I would hope that all the people on this telephone call today...will look very closely at some of the environmental problems in their communities and how many people can be put to work in cleaning those up."[8]

The goal of the confab was to aid government officials, nonprofit organizations, and academicians in their understanding of Agenda 21, plus provide everyone with details on how their participation would be required to fully integrate sustainable development into all aspects of American life.

Fifteen hundred people attended the three-day event, titled "From Rio to the Capitols: State Strategies for Sustainable Development." Predictably, the opening speaker was Vice President Al Gore.

Next, on June 29, 1993, President Clinton signed Executive Order 12852. The order established the president's Sustainable Development Council, which mimicked the United Nation's Council on Sustainable Development. Appropriately, Gore was given authority over the U.S. assembly and would aid in selecting its twenty-five green members, which included:

- Cochairman Jonathan Lash, president, World Resources Institute, an organization dedicated to "putting environmental issues on the international agenda."[9]

- John Adams, executive director, Natural Resources Defense Council; their Mission Statement declares that the "NRDC strives to help create a new way of life for humankind."[10]

- Michelle Perrault, international vice president, Sierra Club, whose goals include "Creating the domestic conditions needed for the U.S. to lead in negotiating and implementing an international climate treaty."[11]

- John Sawhill, president, the Nature Conservancy, an organization that "is working with world leaders to build support of an international climate change agreement," which will include "funding...to help buffer the impacts of climate change on people."[12]

Also on the council were key members from the Clinton administration, who remain active in the environmental movement today.

- Bruce Babbitt, secretary of the interior. Babbitt is now a director of the World Wildlife Fund, a radical, environmental group that is hostile to humankind and seeks "to be the voice for those creatures who have no voice."[13]

- Carol Browner, administrator, Environmental Protection Agency. Browner stepped down as President Obama's director of the Office of Energy and Climate Change Policy in 2011.

- Andrew Cuomo, secretary, Department of Housing and Urban Development. Cuomo is a career politician who is currently the governor of New York. In 2008, as New York's attorney general, Cuomo filed a lawsuit against the George W. Bush administration's EPA for "refusal to control pollution from oil refineries."[14] Cuomo claimed Bush had "a do-nothing policy on global warming."

- William Daley, secretary, Department of Commerce. Daley was also appointed to the board of government mortgage giant Fannie Mae. He went on to chair Al Gore's 2000 presidential campaign. After several lucrative years in the elite banking industry, he became Obama's chief of staff in 2011.

- Bill Richardson, secretary, Department of Energy. Richardson ran for president in 2008, proposing to cut U.S. oil demand by 50 percent by 2020 and mandate a 90 percent reduction on carbon dioxide emissions by 2050.[15]

There were also a few token members of the council representing big business. Among those in this crony clique was Kenneth Lay, CEO of energy-producing giant Enron. Lay was pushing for an international climate treaty in hopes of imposing a cap-and-trade scheme upon American industry. Enron's bean counters had ingeniously crafted an accounting mechanism that would allow the corporation to generate billions of dollars off of such a plan. According to the *Washington Post*, Lay had a special meeting with Clinton and Gore to discuss global warming, and in that meeting "there was broad consensus in favor of an emissions trading system."[16] According to an internal Enron memo, such a trading system would "do more to promote Enron's business than almost any other regulatory initiative outside of restructuring the energy and natural gas industries" and would be "good for Enron stock."[17]

Lay would eventually be indicted on federal charges involving securities fraud in 2004. He was declared guilty in 2006, but mysteriously died shortly before he was to be sentenced. Several other Enron executives were also convicted of similar charges, and the company ended up being dismantled after filing for bankruptcy.

Meantime, the Sustainable Development Council soon cranked out its goals in a document titled *A New Consensus for the Prosperity, Opportunity and a Healthy Environment for the Future*. Key features of the document clearly score well with the ambitions of the United Nations and fly in the face of the foundations of American liberty:

> We Believe, Economic growth, environmental protection, and social equity are linked. We need to develop integrated policies to achieve these national goals.[18]
>
> The United States should have policies and programs that contribute to stabilizing global human population.[19]
>
> Even in the face of scientific uncertainty, society should take reasonable actions to avert risks where the potential harm to human health or the environment is thought to be serious or irreparable.[20]

> The Council should not debate the science of global warming, but should instead focus on the implementation of national and local greenhouse gas reduction policies and activities, and adaptations in the U.S. economy and society that maximize environmental and social benefits, minimize economic impacts, and are consistent with U.S. international agreements.[21]

Now you understand why, to this day, Al Gore repeatedly makes statements like, "There's no legitimate debate,"[22] or that those who do not believe in the theory of human-caused global warming are "like the ones who still believe that the moon landing was staged in a movie lot in Arizona and those who believe the earth is flat."[23]

Gore's *A New Consensus* also states, "The federal government should play a more active role in building consensus on difficult issues."[24]

Science is not conducted on the basis of consensus—never has been and never should be; and in this case, "building consensus" implies serious arm-twisting. Such methodology leads to reckless decisions based on the input of junk science, emotion, and the desire for devious agendas.

According to the National Academy of Science, this notion of building consensus is contrary to the skepticism that is required to conduct an honest debate: "Scientific knowledge and scientific methods, whether old or new, must be continually scrutinized for possible errors. Such skepticism can conflict with other important features of science, such as the need for creativity and for conviction in arguing a given position."[25]

ECOSYSTEM: IN THE EYE OF THE BEHOLDER

Simultaneous to the creation of the Sustainability Council, Al Gore was ready to check off another box on the Agenda 21 to-do list: restructure the governmental decision-making process. According to the Agenda, "Governments should conduct a national review and, where appropriate, improve the processes of decision-making so as to achieve the progressive integration of economic, social and environmental issues in the pursuit of development that is economically efficient, socially equitable and responsible and environmentally sound."[26]

President Clinton allowed Gore to conduct a "National Performance Review" (NPR). The general public was made to believe the goal of the

NPR was to streamline government bureaucracy; indeed, as a part of the NPR, Gore even dreamed up an award to recognize federal employees who supposedly engaged in "reinventing government." The accolade was coincidentally called the "Hammer Award" (Gore never would say if his inspiration for naming the award was related to a *certain* dear friend of the family). However, despite the window dressing, the NPR's primary achievement was to enable the Department of Interior, Agriculture, the National Oceanic and Atmospheric Administration, and the EPA to work in concert with one another to meet, according to Gore, the "goals of continued vigorous economic growth and preservation of our magnificent natural heritage."[27]

The vice president was essentially making good on the belief that "economic growth, environmental protection, and social equity are linked." To further this effort, Gore instituted the Ecosystem Management Initiative, which empowered government assets to define, protect, and manage America's ecosystems. However, many critics of the plan immediately asked an obvious question: what is an ecosystem? An ecosystem can be as small as a backyard or as large as the Mississippi River drainage basin. A pond can be an ecosystem, as are the oceans. The territory that forms the habitat of a single eagle, or the region that is shared by two species of trees, can be considered an ecosystem. Government policy aimed at protecting ecosystems is a wide-open opportunity for draconian regulations—which is why environmentalists immediately loved Gore's plan.

Illustrating my point, during a congressional hearing, Sustainable Development Council member Bruce Babbitt was asked by representative Jay Dickey (a Republican from Arkansas) to define an ecosystem:

> **Mr. Dickey**: Good morning, Mr. Secretary. In your mind, what is an ecosystem, how will one be defined, and how will you differentiate one from another?
>
> **Secretary Babbitt**: Mr. Congressman, to some degree, an ecosystem is in the eye of the beholder.
>
> **Mr. Dickey**: Is that your answer? Would you like to elaborate?
>
> **Secretary Babbitt**: I think I would be willing to elaborate, sir. I can put it in specific context. The timber problem and the salmon problem

> drives you to an ecosystem which essentially runs from the crest of the Cascades to the Pacific Ocean from approximately Puget Sound to the beginning of the Sierra Nevada in California. It is characterized by a lot of the commonalities. Stream drainage is certainly a big one.
>
> Climate. The weather from the Pacific creates a lot of precipitation until it hits the tops of the Cascades and then you are off into the desert. So that is an ecosystem. River basins are a pretty good starting point.
>
> In the case of the Edwards Aquifer in Texas, we were looking at an ecosystem defined by ground water recharge in the limestone hill country of west Texas. In other cases, the dominant thing will be the vegetation communities. Some would say the Colorado Plateau is an ecosystem. Others would vigorously dissent.
>
> I think the essential thing you are looking for is common natural and geographic features that generate a particular set of resources or a particular set of problems or opportunities.[28]

Babbitt's admission that an ecosystem is "in the eye of the beholder" was staggering, and threw open the barn door for government to run wild in its effort to implement the goals of Agenda 21. The Ecosystem Management Initiative has since allowed the federal government more power to legally and more efficiently regulate land use in, and, when necessary, take land from the property owner, in spite of the Fifth Amendment to the Constitution.

The Fifth Amendment requires just compensation for land taken by the federal government. However, liberal interpreters of the Constitution see the federal protection of ecosystems as offering a way to circumvent the Fifth Amendment by declaring ecosystems entities that belong to the public and that therefore *must* be protected under laws regarding the public trust. The doctrine of public trust ensures that there are responsibilities vested in government that cannot be relinquished; thus, it is the duty of government to protect the people's common interest in certain resources.

One of the first publicized examples of this taking of land in the name of an ecosystem involved the family of Mr. William Stamp in Rhode Island. The Stamp family has farmed in Rhode Island for several generations. Because of the urbanization of the region surrounding their farmland, their property values were reassessed and the associated property taxes rose from four thousand dollars to seventy-two thousand dollars per year. To offset their tax burden, the Stamps tried to sell seventy acres to developers. Throughout Clinton and Gore's eight years in office, the sale was prevented because the parcel in question had been determined—"in the eye of the beholder"—to be a special ecosystem, in this case a wetland. Since wetlands are protected by the federal Clean Water Act, the Stamps had to find seventy-two thousand dollars annually to stay above the law. After years of turmoil, the sale of the Stamps' private property was finally remedied when clearer eyes came into focus during the subsequent Bush administration. However, as we will later discuss, the Obama administration is pulling out all the stops to play the ecosystem card whenever necessary to prevent development and usurp property rights.

GORE'S GROWTH BOOK

Another lasting Gore legacy is a manual whose creation was encouraged by the Sustainable Development Council. The *Growing Smart Legislative Guidebook: Model Statutes for Planning and the Management of Change* was created with funding from the Department of Housing and Urban Development.[29] It is the most significant environmentally based bureaucratic compendium in use today. There is not a planning commission at any level of government in the United States that is not familiar with this book. *Growing Smart* literally provides intricate details on how to create environmental policies, regulations, laws, and even punishments for those guilty of violations. Gore's influential growth book fulfills the wishes of yet another Agenda 21 goal:

> Laws and regulations...are among the most important instruments for transforming environment and development policies into action.[30]
>
> Each country should develop integrated strategies to maximize compliance with its laws and regulations relating to sustainable development...
>
> The strategies could include, enforceable, effective laws, regulations and standards that are based on sound economic, social and environmental principles and appropriate risk assessment, incorpo-

> rating sanctions designed to punish violations, obtain redress and deter future violations.[31]

If a local planning department wants to prevent a shopping center being built, *Growing Smart* has the solution. If the public needs to be convinced that those trees on the other side of town must be preserved, the guidebook has the answers. If a punishment needs to be prescribed for those who breach environmentally based laws, the growth book has all the answers.

Confirming how influential the president's Sustainable Development Council was, in 1997, a little-known document was prepared by the Clinton administration for the United Nations, titled "Implementation of Agenda 21: Review of the Progress Made Since the United Nations Conference on Environment and Development, 1992." In it we read that the U.S. government has pledged to be the vanguard of the international Green Agenda:

> The U.S. Government remains committed to promoting sustainable development... The U.S. believes that the CSD [United Nations' Council on Sustainable Development] should continue to serve as a focal point for monitoring the implementation of Agenda 21 at local, regional and international levels.
>
> The President's Council on Sustainable Development (PCSD) is the key national sustainable development coordinating mechanism of the United States...[and] continues to provide a broad mandate for federal agencies to create and maintain conditions under which man and nature can exist in productive harmony and fulfill the social, economic and other requirements of present and future generations of Americans.
>
> The U.S. Government promotes policies and programs in the areas of energy efficiency, environmentally sound and efficient transportation, industrial pollution control, sound land-use practices, sound management of marine resources and management of toxic and other hazardous waste.[32]

President Clinton's Council on Sustainable Development was a seditious cabal of anti-American operatives working at the behest of the United Nations. And all of its activities were overseen and approved by Al Gore.

FINAL NAIL IN THE COFFIN

By the end of the Clinton-Gore administration, the goal of utilizing the environment to undermine America's system of free enterprise, capitalism, and the right to one's own property, had made significant advances. However, there was one critical piece that needed to be accomplished. Carbon dioxide would need to be deemed an official pollutant.

In the next chapter I will provide you with a detailed examination of carbon dioxide's role in our atmosphere, but let me just state that carbon dioxide, CO_2, is a fertilizer, not a pollutant.

The United States has done a remarkable job of cleaning up the pollution within its skies. I recall that as a young child living in the "smog capital" of the country, Los Angeles, in the sixties, my lungs would burn while I was playing football in the front yard. Indeed, back then L.A. would issue alerts for dangerous levels of smog more than two hundred days per year. Now, despite more people, cars, buses, and planes, Los Angeles issues fewer than a dozen such alerts annually—and the standards for such warnings have actually become stricter.

Nationally, statistics compiled by the American Enterprise Institute reveal that in the twenty-five years spanning 1980 and 2005, the pollutants named in the 1970 Clean Air Act were reduced markedly:[33]

- Fine particulate matter declined 40 percent.
- Ozone levels declined 20 percent, and days per year exceeding the eight-hour ozone standard fell 79 percent.
- Nitrogen dioxide levels decreased 37 percent, sulfur dioxide dropped 63 percent, and carbon monoxide concentrations were reduced by 74 percent.
- Lead levels were lowered by 96 percent.

What makes these air quality improvements even more noteworthy is that they occurred during a period in which:

- automobile miles driven each year nearly doubled to 93 percent, and diesel truck miles more than doubled to 112 percent

- tons of coal burned for electricity production increased over 60 percent

- the real dollar value of goods and services (gross domestic product) more than doubled to 114 percent

So, given that bona fide pollution had been cleaned up in America, the environmentalists needed to create a new pollutant, in this case a vital component of the atmosphere that no one is able to touch, taste, see, or smell: CO_2. Gore and his minions wanted the EPA to be able to regulate carbon dioxide under the Federal Clean Air Act.

It took a few years, but finally the left achieved a monumental victory on April 2, 2007. That was the date when the Supreme Court threw science under the bus by declaring that CO_2 was a *pollutant.* The petitioners were the states of California, Connecticut, Illinois, Maine, Massachusetts, New Jersey, New Mexico, New York, Oregon, Rhode Island, Vermont, and Washington, plus the cities of New York, Baltimore, and Washington, D.C., plus the territory of American Samoa, of all places. Eco-socialist organizations including the Friends of the Earth, Greenpeace, Natural Resources Defense Council, and the Sierra Club were also involved.

The Court's landmark 5–4 decision required the EPA to regulate CO_2 under the Clean Air Act. President Bush refused to allow his EPA to respond to the court ruling, but within his first year on the job, predictably, President Barack Obama did.

December 7, 2009, lived up to its 1941 billing as the day that will live in infamy as Obama's EPA chief, Lisa Jackson, held a press conference to announce "that the EPA has finalized its endangerment finding on greenhouse gas pollution, and is now authorized and obligated to take reasonable efforts to reduce greenhouse pollutants under the Clean Air Act." The "endangerment finding" means humanity is at risk because of a miniscule increase in carbon dioxide, and the Obama administration believes it has the dutiful power to react by curtailing such emissions.

"The overwhelming amounts of scientific study show that the threat is real—as does the evidence before our very eyes," Jackson said. "Polar ice caps crumbling into the oceans, changing migratory patterns of animals and broader ranges for deadly diseases, historic droughts, more powerful storms, and disappearing coastlines. After decades of this mounting evidence, climate change has now become a household issue. Parents across the United States and around the world are concerned for their children and grandchildren."

Taken to an extreme, virtually anything that emits carbon dioxide could be regulated by the EPA—power plants, planes, cars, lawn mowers, even marathon runners. That's why it was no surprise to me when in August 2011, it was announced that the EPA will propose standards on power plants and oil refineries to clamp down directly on the greenhouse gases they claim are responsible for global warming, possibly forcing up to one-fifth of the nation's oldest coal-fired electrical generating plants, largely in the Midwest and South, to shut down in the next five years. Given that coal provides nearly 50 percent of the nation's power, that means higher electric bills, rolling blackouts, and fewer jobs.

GOREBASMS

So how does Al Gore respond to anyone willing to think for himself and dig into the data to form his own opinion about global warming—which, by the way, usually results in that person becoming a skeptic or denier of anthropogenic global warming? Gore reacts with bizarre "Gorebasms."

Gorebasms are pejorative statements directed toward deniers and skeptics. Al's been doing this for years, like the time in 2007 when he went after Dr. John Christie, a Nobel laureate and top-shelf researcher from the Earth System Science Center at the University of Alabama. Christie is one of the people responsible for collecting the earth's temperature data via satellite and is *not* in Gore's camp. Dr. Christie wrote an opinion piece published by the *Wall Street Journal* stating, "I see neither the developing catastrophe nor the smoking gun proving that human activity is to blame for most of the activity we see."[34]

Soon after the article ran, the former vice president was being interviewed by Meredith Vieira on NBC's *Today Show*. Vieira brought up the op-ed to which Gore "basmed" that Christie was "way outside the scientific consensus." He also sharply criticized the media for giving so much airtime to such climate skeptics, whining, "But, Meredith, part of the challenge the news media has had in covering this story is the old habit of taking the 'on the one hand, on the other hand' approach. There are still people who believe that the earth is flat, but when you're reporting on a story like the one you're covering today, where you have people all around the world, you don't take—you don't search out for someone who still believes the earth is flat and give them equal time."[35]

Since then, hundreds of noteworthy scientists have come out against Gore's position. One of the twelve people who walked on the moon, highly acclaimed geologist Dr. Harold Schmidt, said it was "ridiculous" to claim there is a "consensus" of scientists who believe "that humans are causing global warming when human experience, geological data and history, and current cooling can argue otherwise."[36]

The world's first woman to receive a PhD in meteorology, Joanne Simpson, formerly of NASA, has also voiced her concerns, stating, "Since I am no longer affiliated with any organization nor receiving any funding, I can speak quite frankly... As a scientist I remain skeptical."[37]

Another big blow to the theory came from Harold Lewis, emeritus professor of physics at the University of California, Santa Barbara, who proclaimed, "Global warming is the greatest and most successful pseudo-scientific fraud I have seen in my long life."[38]

The most recent thumb in the eye of Al came from Dr. Ivar Giaever, a Nobel Prize winner for Physics. Giaever was one of seventy Nobel laureates who signed an open letter endorsing Barack Obama for president in 2008. On September 13, 2011, he wrote a letter to the American Physical Society (APS) presenting the reason for not renewing his membership with the organization:

> I did not renew it because I cannot live with the [APS] statement below [on global warming]:
>
> > The evidence is incontrovertible. Global warming is occurring. If no mitigating actions are taken, significant disruptions in the Earth's physical and ecological systems, security and human health are likely to occur. We must reduce emissions of carbon dioxide now.

Dr. Giaever went on to state that in his opinion the planet's temperature is, in fact, "amazingly stable."

All of these brilliant minds coming out against the Green Agenda's premier theory have been too much for Gore, and now he's clearly becoming mentally unstable. During a speech turned wild rant before a gathering of elites at the Aspen Institute in August 2011, Gore claimed special interest groups "pay pseudo-scientists to pretend to be scientists to put out the message: 'This climate thing, it's nonsense. Man-made CO_2 doesn't trap heat. It may be volcanoes.' Bullshit! 'It may be sun spots.' Bullshit! 'It's not getting warmer.' Bullshit!" It was a gorebasm in which Al totally lost it. Listening to audiotape of the speech makes it abundantly evident even the Aspen audience was uncomfortable witnessing the ravings of a madman.

And just in case there might be someone present with a view contrary to his, Al intimidated the crowd, blustering, "When you go and talk to any audience about climate, you hear them washing back at you the same crap over and over and over again. They have polluted the shit. There's no longer a shared reality on an issue like climate even though the very existence of our civilization is threatened. People have no idea!"[39]

My heart goes out to you, Tipper. You certainly deserve better than this buffoon. I hope you've received a big cut from the divorce.

And there is a lot of green—as in cash—to be divided.

GORE'S BILLIONS

With initial thanks to my Silicon Valley friend, Dave, I've learned quite a bit about how Gore has made, is making, and stands to make, unfathomable amounts of money off the global warming scam. Keep in mind we're talking about a man who, according to disclosures provided to the media during his run for the White House in 2000, was worth a couple million. Following that failed bid he "cofounded" the financial investment firm Metropolitan West with Philip Murphy, formerly of Goldman Sachs (GS) and previously the Democratic finance committee chairman. I emphasize with sarcasm "cofounded."

Prior to 2000, Gore's financial experience included a dubious visit to a Chinese Buddhist temple in California that yielded thousands of dollars in illegal campaign contributions. As *Fortune* magazine reported, Gore's visit to the temple was "the very symbol of campaign-finance chicanery."[40]

In 2001, Gore was recruited as an adviser to Google, scoring loads of preferred stock. Those shares are now worth untold millions.

In 2003, Gore's net worth shot even higher as he took a position on the board of Apple, securing both an undisclosed director's salary and significant shares of stock. At the time, Apple was trading at $21 per share. As of this writing, Apple is trading at over $370. Al was given another home run in 2003 when Metropolitan West sold to Wachovia Bank for an estimated $50 million.

Next Murphy put Gore together with David Blood, another Goldman Sachs partner, and they founded Generation Investment Management (GIM) in London. GIM is a hedge fund that invests in companies supposedly committed to "going green." Founding partners include Goldman CEO and U.S. Treasury secretary under George W. Bush, Hank Paulson. Based upon twenty-nine known holdings in the GIM portfolio, it's estimated that the fund is valued at $3.3 billion—three times up from the amount I reported in *Climategate* in 2010.[41]

Besides Google and Apple, Gore's extensive connections to Silicon Valley include a partnership in the foremost venture capital firm in the Bay Area, Kleiner Perkins (KP). KP is run by Al's longtime pal John Doerr. Among KP's hits are early financial backings for Google, Netscape, Sun, and Amazon. Upon joining KP, Gore claimed that he would give 100 percent of his KP salary to his public relations firm, the Alliance for Climate Protection, which is run as a nonprofit entity. However, Gore's giveaway is likely a total sham, because he's entitled to a tax write-off for the charitable contribution *and* is legally entitled to receive both a salary from the nonprofit and paid expenses.

In 2008, Gore and Doerr started a $500 million "Green Growth" fund, hiring a team from Goldman Sachs' Special Situation's Group to hunt for deals involving start-ups and young companies trying to make a play in the alternative energy field.[42] The fund, which is a mighty arm of the Kleiner Perkins portfolio, is thought to be worth close to a billion dollars.[43] The fund also invested $75 million into Silver Spring Networks, a company that provides essential systems infrastructure for the smart grid (to be discussed in chapter 12).[44] And the same year, Gore reportedly invested $35 million with Capricorn Investment Group LLC, a Silicon Valley firm that, according to a filing with the Securities Exchange Commission, selects the private funds for clients and invests in makers of environmentally friendly products.[45] Capricorn was cofounded by billionaire Jeffrey Skoll, former president of eBay and, conveniently, the executive producer of Gore's Oscar-winning film. Another Capricorn cofounder, Ion Yadigaroglu, described the firm's aggressive business model, stating, "We're trying to make more money than others doing the same thing and do it in a way that is superior."[46]

To keep the pockets of Gore and friends stuffed with cash, there has to be global warming. And that's why deniers and skeptics are maligned, marginalized, and gorebasmed.

A BATTLE WON

Climategate was released on Sean Hannity's television show on the Fox News Channel on Earth Day 2010. On the program, I stated that the reason I wrote the book was to "arm Americans with the truth in order to keep Nancy Pelosi and Harry Reid from passing a cap-and-trade scheme that would harm businesses and cause the price of virtually everything to go up."

It worked. Try as they might, Pelosi and Reid were unable to push through cap-and-trade, even though they held majorities in their respective chambers. The political climate was shifting, and passing such a bill would have doubled the resentment the Democrats were already experiencing from constituents peeved about the Obamacare law. Certainly the loss of cap-and-trade made Al Gore and a host of others livid. Not only would it have ushered in the UN dream of altering U.S. lifestyles, transferring wealth, and diminishing property rights, but cap-and-trade would have been like printing money for those—like Gore—who were properly invested. It's just one of the reasons Gore is showing signs of losing it.

Another cause for Gore's indignation is that the earth's temperature is *not* warming...

FIVE

GLOBAL WHINING

FOR YEARS I'VE OWNED the Internet URL DebateMeAlGore.com. I use the provocative address to direct people to the website where I post my environmental blogs. Al Gore certainly knows who I am, but the likelihood of the two of us debating is nil. Gore and his kind don't want to debate anyone regarding the theory of anthropogenic global warming—because they realize the theory is sorely flawed. It's just another reason why Al is cussing mad.

Let's begin with the theory:

> The earth's atmosphere is warming as a result of an increase in greenhouse gases generated by human activity, particularly through the prolific use of fossil fuels and their byproduct, carbon dioxide.

The theory of global warming sounds intriguing. After all, we're told the weather is getting warmer, and it's apparent there are more people using more fossil fuels than ever. We know that the by-product of fossil fuel is carbon dioxide (CO_2), and we're also told that atmospheric levels of carbon dioxide are increasing, so the experts *must* be correct, and therefore, the theory *must* be true.

Problem is, the theory is patently false and needs to be thrown in the heap adjacent Ptolemy's hypothesis that the sun and planets revolve around Earth.

THE HISTORICAL RECORD

Until 1990, it was generally accepted by all corners of the scientific community that over the past one thousand years, the temperature record of the earth illustrated a distinct warming period between AD 900 and 1300. During this four-hundred-year span, the average temperature on the earth's surface was likely at least 2 degrees (Fahrenheit; 1.2 degrees Celsius) warmer than today. This period is known as the Medieval Warm Period (MWP). That same body of scientists also agreed that from AD 1350 to 1800 there was a global cooldown, reversing the warmth of the MWP, with temperatures at one point falling to roughly 2 degrees (1.2 degrees Celsius) cooler than today. This 450-year period is known as the Little Ice Age (LIA). Following the LIA, temperatures stabilized for approximately 50 years.

Then, between 1850 and 1940, temperatures increased 1 degree (.60 degree Celsius).

Next, from 1940 to 1970, the earth cooled about .20 degree (.10 degree Celsius); this was followed by a minor warming of a mere .34 degree (.20 degree Celsius) occurring between 1970 and 1998.

Thus, since 1850 there has been an increase in temperature of 1.14 degrees (.70 degree Celsius), and 88 percent of this warming occurred *before* 1940. These are facts that are indisputable. And, it's also a known fact that since 1998 there has been no additional warming. Period.

However, let's pause here to talk about the radical revisionism in temperature record keeping that has been occurring since 1998.

A true founding father of modern meteorology is Dr. Hubert H. Lamb. In 1972, Dr. Lamb founded the Climatic Research Unit at East Anglia in the United Kingdom (CRU). The work conducted at CRU was noted to be the best in the world, and confirmed the thousand-year temperature record noted earlier. In fact, a graph representing those ebbs and flows was created by Lamb and used by the United Nations in all of its climate documents for years—including in its much revered 1990 International Panel on Climate Change report (IPCC).

However, environmentalists would always become unglued when inquisitive minds would ask, "So what caused the temperature to climb during the Medieval Warm Period? It certainly couldn't have been the SUV. Is it possible that temperatures can rise and fall *naturally*?"

These constant questions became a source of great irritation for those who desired to see the world's economy being based on the principle of sustainable development. Something needed to be done. A new graph would have to be produced, but quietly. In fact, it would be best if the graph could be procured behind Dr. Lamb's back. You see, even though Dr. Lamb founded CRU, he stepped down as its director in 1978, but remained on staff as an instructor, researcher, and mentor—and he was not a fan of the new anthropogenic theory.

Lamb remained employed by CRU until his untimely death in 1997. As a reminder of where he stood on the subject, an excerpt from his obituary, which ran in the *Independent* in London on Wednesday, July 9, 1997, clearly states:

> An irony is, now that the world is acutely aware of global climate change, Lamb had maintained a guarded attitude to the importance of greenhouse gas warming. Although many others have accepted this, he felt that there was too much reluctance to consider the full range of other, natural, causes of change. Right to the end of his life, he was promoting his "different view."
>
> His different view of climate has left behind a deeper understanding of the nature of climate change, and of the interactions between natural systems which contribute to it.

Cruel as it sounds, with Lamb out of the way, the CRU could now trash his graph and contrive a version to make it appear as if the planet were experiencing the hottest climate swing in history. The assignment went to a close associate of CRU, Michael Mann, who at the time worked in the University of Massachusetts' Department of Geosciences. To arrive at his conclusions, Mann cleverly used much of the same data previously inputted for the Lamb graph, including highly respected tree ring research as his primary data to account for the medieval years.[1] For more recent centuries, Mann inserted measurements from sea sediment, ice core samples, and oxygen isotopes—all sound, legitimate sources. However, Mann's process began waxing weird when he threw into the mix temperatures logged in major cities that have undergone huge artificially induced upward swings due to the heat-trapping influences of concrete, asphalt, and steel (known as the urban heat island effect). Stranger still, he conveniently excluded the highly accurate temperature record recorded by flawless satellite instrumentation. In addition, Mann did not discard wild temperature anomalies associated with the mysterious alteration of the Pacific Ocean's currents, known as El Niño, which created a significant global warmup from 1982 to 1983 and 1997 to 1998.

In the end, Mann had meticulously manipulated the climatic record. The warmth of the MWP and chill of the LIA had been smoothed out to make the bulk of the past millennium rather homogenous. The result was like an ice pick to the climate's prefrontal cortex: the past had been forgotten. Mann's work was presented to the world in a 1998 paper published in *Nature*,[2] with another presented in *Geophysical Research Letters*.[3] For those seeking to use science to achieve political and social change via environmental policy, the papers were like proverbial manna from heaven.

The final object of Mann's climatic memory-buster was a bogus graph that looked like a hockey stick positioned horizontally, with the blade protruding straight up. In fact, the chart became known as "Mann's hockey stick" and became a popular device to convince the uninformed observer that the earth has a fever.

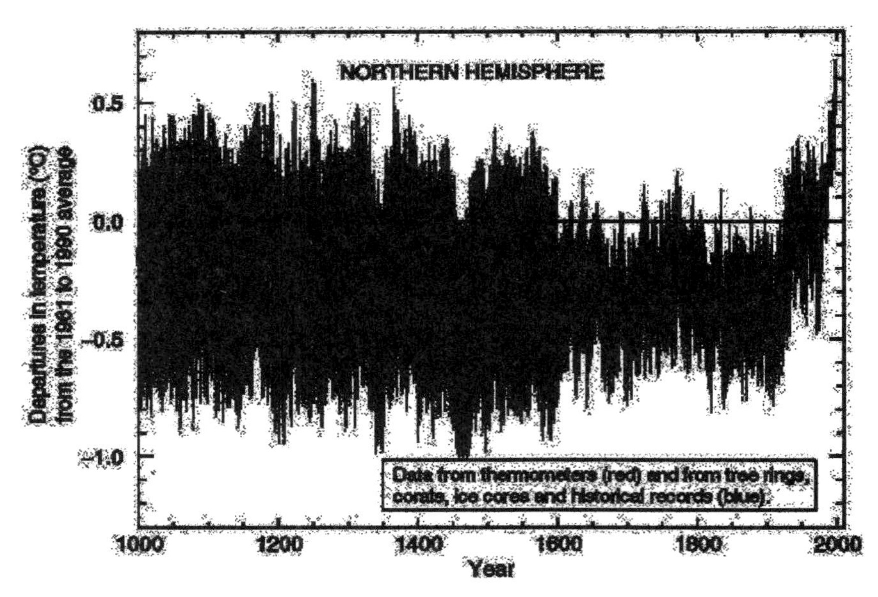

Mann's Hockey Stick, from the International Panel on Climate Change Third Assessment Report, 2001[4]

Observing Mann's graph, one can clearly see a dramatized temperature spike ensuing about 1930 and reaching warp speed near 1970. Also, note that the Medieval Warm Period, and its counterpart, the Little Ice Age, have been expunged from the record.

Despite a plethora of scientists with PhDs coming forward to bust Mann's hockey stick, this horrendous bastardization of science remains a CRU-sponsored icon. The United Nations uses the graph to compel global policy, while Al Gore carries it to the bank, as the contrivance is a pillar in his media presentations.

And those who dare criticize the spurious Stick are branded "outliers."

HOTTEST WEATHER EVER?

So how do we win the argument that this is *not* the hottest weather ever?

We do it with the facts.

To begin with, we've only possessed the ability to precisely measure the temperature with thermometers since the early 1800s, just a few decades before the conclusion of the Little Ice Age and the beginning of the Industrial Revolution (1850). Since then, according to the National Climatic Data Center, the warmest decade on record in the United States was the 1930s, with twenty-two of the now fifty states recording their highest temperatures ever during those years. Examples include 110 degrees in Millsboro, Delaware, on July 21, 1930; 118 degrees at Keokuk, Iowa, on July 20, 1942; and Steele, North Dakota, reaching 121 degrees on July 6, 1936. Thirty-eight states recorded their all-time highs before 1960. Likewise, the hottest year on record in the U.S. was 1934. Since our nation has kept the most complete, widespread, long-standing, uninterrupted set of temperature records on the planet, these records should not be taken lightly.

Second, most of the world's thermometers have only been installed since the 1980s, including more than a thousand in the U.S., with four hundred of those placed after 1998. Logically, with the additional data points entered from locations never before measured, there is a statistical likelihood that the overall average temperature of the planet could skew warmer. That's especially possible when you realize that the man responsible for much of the planet's thermometer distribution and monitoring is James Hansen, director of the Goddard Institute for Space Studies, a division of NASA. You will learn more about Hansen, and his wild rhetoric, in the next chapter, but allow me to introduce you to him via a brief remark he made in 2009: "We have only four years left."

By the way, if the designated record keepers would rely solely on satellite data, which accurately monitors the entire planet and has been providing readings since 1979, they would see figures confirming that the planet's temperature is not warming. In fact, Phil Jones, who worked for Dr. Lamb at CRU and eventually became its director in 1998, finally admitted in 2010 that over the last fifteen years there has been no "statistically significant" warming.[5]

Nonetheless, year after year the National Oceanic and Atmospheric Administration (NOAA) cranks out variations of the same press release declaring this to be the hottest, or near hottest, weather on record. The first line from their January 2011 statement declares, "According to NOAA scientists, 2010 tied with 2005 as the warmest year of the global surface temperature record, beginning in 1880."

The problem is the entire temperature record has been wrecked by Hansen, et al. An excellent 110-page paper on this bastardization of science was prepared by a man who for years served as the Weather Channel's chief meteorologist, Joe D'Aleo, and the web's most tenacious atmospheric blogger, meteorologist Anthony Watts of surfacestations.org. In summary they wrote:

- NASA and NOAA (a.k.a. "the federal government") are now inputting less (not more) temperature reporting stations in determining the global average (they used data from nearly 6,000 stations in 1970, and rely on about 1,500 today).

- Among reporting stations omitted are those in higher latitudes, higher elevations and rural locations (away from the artificial warming found in cities).

- The feds are "correcting" temperatures from older records downward, to create the impression of a current warming trend.

- They are not properly compensating for urban growth and land-use changes that can produce localized warming known as the urban heat island effect. The feds are essentially cherry-picking thermometers from reporting stations sited at busy airports and other artificially warm locales.

- NOAA now collects data from only 35 sites in Canada, down from 600 in the 1970s.

- After 1990, NOAA tripled the number of Canadian reporting stations at lower elevations while reducing by half the number at elevations above 300 feet.

- Globally, reporting stations in the Andes and Bolivia have been omitted from the record.

- Only 25 percent of Russia's reporting stations are now included in the global temperature calculations.[6]

There is clearly a deficiency in the way temperatures are currently being collected. However, again, even considering the bias in recent years, the earth's climate has only warmed .7 degree Celsius since 1850—with most of that occurring before 1940. Again in 1850, the planet was un-thawing from the long, frigid Little Ice Age, so some warming since then would be expected. But the global warmers don't really want you to observe the hard data; instead they demand you look at yourself. During those sixteen decades, the human population has increased from 1.1 billion to about 7 billion; we've gone from no gas-powered vehicles to roughly 1 billion;[7] and we've seen the proliferation of the train, plane, natural gas to heat homes, and coal to create electricity (all of which emit carbon dioxide). *You* are the problem, they contend, even though all they have to show for it is a paltry seven-tenths of a degree rise in temperature, with most of it occuring seventy years ago, and no sign of warming since at least 1998.

For those searching for cause and effect, the relationship between human activity and temperature is pathetically unconvincing.

BUSTING MYTHS

For example, let's quickly examine some of the popular icons that supposedly "prove" there is global warming:

- SEA LEVEL RISE: It's a fact that ever since the end of the last Ice Age, global sea level has been gradually increasing. The melting ice and snow from that bitterly cold event is continually trickling into our great oceans and seas. According to the IPCC, over the past twenty thousand years, sea level has increased nearly four hundred feet.[8] While that may sound like a lot, it equals a rise of merely 1.8 millimeters per year.[9] Try placing your thumb as close to your forefinger as possible, without the two touching—that's how much the sea has been rising each year, for hundreds of years. Hardly frightening.

- SINKING PACIFIC ISLANDS: For years, Al Gore and the greens have been telling us that the South Asia island chain, the Maldives, and Tuvalu, halfway between Hawaii and Australia, are sinking. Not true. The Maldives are relatively flat atolls, composed of coral. Since 1972, tourism has become the central focus of the Maldivian economy, with the number of resorts zooming from one to eighty-eight today, jammed primarily atop three islands: the North and South Malé Atolls and Ari Atoll. Locally mined coral rock has been the main aggregate for constructing these resorts.[10] The mining has severely compromised the atolls, creating the impression that the islands are sinking.

  Tuvalu's problem is that it's a country that was never meant for modern habitation. Their primary indigenous vegetable crop, taro, has been gravely over-farmed. There is no fresh water available—only what can be cached from rain. Much of the population on the main island use a lagoon for their bathing and toilet facilities. The tiny country ships its commercial waste to landfills in Fiji and New Zealand. As for the sea level, according to research conducted by the Tuvalu Meteorological Service, the ocean waters about Tuvalu are in *decline*.[11] Confirming this, Professor Patrick Nunn, who researches sea level at the University of the South Pacific in Fiji, admitted, "A lot of these sea gauges have been slowly falling over the last five years."[12] The actual decrease appears to be 2.5 inches.[13] Likewise, according to a paper published in *Global and Planetary Change*, Stockholm University professor Nils-Axel Mörner says, "We were unable to detect any traces of a recent sea level rise [around the Maldives]. On the contrary, we found quite clear morphological indications of a recent fall in sea level."[14] His research indicated that sea level about the Maldives has fallen approximately eleven inches in the past fifty years.

- MELTING ARCTIC SEA ICE: Sea ice has only been scientifically monitored since the introduction of satellite technology in 1979. So, when you hear reports of "the least amount of ice in history," the history they speak of is just slightly over thirty years. Truth is, naval records illustrate that in the 1950s there were periods when the waters near the North Pole were virtually ice-free. In a comment originally posted by the late global warming blogger John Daly, a former submariner stationed aboard the USS *Skate* said that in 1959 "the *Skate* found open water both in the summer and following winter. We surfaced near the North Pole in the winter through thin ice less than two feet thick."[15] Certainly in 1959 no one was talking about global warming; in fact, the earth was experiencing a cooling. It's more likely that rapid declines in Arctic ice have to do with altered ocean currents and wind patterns, as indicated by research conducted by the Japan Agency for Marine-Earth Science and Technology and published by the journal *Geophysical Research Letters*.[16]

- KILIMANJARO MELTING: At 19,340 feet, Africa's mount Kilimanjaro is a grand sham used by environmentalists eager to play the junk science card. Yes, overall the glacier atop Kilimanjaro is receding; however, it's been in such a state off and on since the end of the Little Ice Age. In November 2007, Britain's *Guardian* ran a story that boldly began, "A new study on the dwindling ice cap of Africa's highest peak, Mount Kilimanjaro, suggests that global warming has nothing to do with the alarming loss of its beautiful snows."[17] The study, published in the *American Scientist*, verifies local legend that the melting of Kilimanjaro was noticed a few decades after the conclusion of the LIA. The research also stated that, while Kilimanjaro's glacier had declined 90 percent between 1880 and 2003, "the decline in Kilimanjaro's ice has been going on for more than a century and that most of it occurred before 1953."

Furthermore, to blame the shrinkage on warmer temperature is shortsighted. According to the study: "Another important observation is that the air temperatures measured at the altitude of the glaciers and ice cap on Kilimanjaro are almost always substantially below freezing (rarely above minus 3°F). Thus, the air by itself cannot warm ice to melting... When pieced together, these disparate lines of evidence do not suggest that any warming at Kilimanjaro's summit has been large enough to explain the disappearance of most of its ice, either during the whole 20th century or during the best-measured period, the last 25 years."[18]

- MORE HURRICANES: While hurricanes are no laughing matter, it's a hoot to hear the global warmers try to associate the grand dames of earth's natural weather machine to climate change. Like all kinds of severe weather, hurricanes simply happen. On average, close to seven hurricanes every four years (1.8 per year) strike the United States, while about two major hurricanes cross the U.S. coast every *three* years. Consider some other noteworthy hurricanes, *none* of which occurred during the Mann hockey stick years:

Deadliest hurricane: More than 8,000 people perished September 8, 1900, when a Category 4 hurricane barreled into Galveston, Texas. The storm surges exceeded 15 feet and winds howled at 130 miles per hour, destroying more than half of the city's homes.

Most intense hurricane: An unnamed storm slammed into the Florida Keys on Labor Day 1935. Researchers estimated that sustained winds reached 150 to 200 miles per hour with higher gusts. The storm killed an estimated 408 people.

Greatest storm surge: In 1969, Hurricane Camille produced a 25-foot storm surge in Mississippi. Camille, a Category 5 storm, was the strongest storm of any kind to ever strike mainland America. When the eye hit Mississippi, winds gusted up to 200 miles per hour. The hurricane caused the deaths of 143 people along the coast from Alabama into Louisiana, and led to another 113 deaths as the weakening storm moved inland.

Earliest and latest hurricanes: The hurricane season is defined as June 1 through November 30. The earliest observed hurricane in the Atlantic was on March 7, 1908, while the latest observed hurricane was on December 31, 1954. The earliest hurricane to strike the United States was Alma, which struck northwest Florida on June 9, 1966. The latest hurricane to strike the United States was on November 30, 1925, near Tampa, Florida.

NATURE'S FERTILIZER: CO_2

But the greatest myth of all is that carbon dioxide is a pollutant. It's astounding to note that, of the gases in our atmosphere, the amount of CO_2, or carbon dioxide, is almost imperceptible. While the primary gases, nitrogen and oxygen, account for 78 percent and 21 percent respectively, CO_2 makes up only .038 percent; that is, a mere 38/1,000ths of a percent of all atmospheric gases.

To better describe this measurement, in *Climategate* I allude to an analogy originally penned by the late Michael Crichton in his must-read eco-thriller, *State of Fear.* On page 387, he likens the gases of the earth's atmosphere to a football field:

> The goal line to the 78 yard-line contains nothing but nitrogen. Oxygen fills the next 21 yards, stretching to the 99 yard-line. The final yard, except for four inches, is argon, a wonderfully mysterious inert gas useful for putting out electronic fires. Three of the remaining four inches are crammed with a variety of minor, but essential, gases. And the last inch? Carbon dioxide. The equivalent of one inch out of a hundred-yard field! If you were in the stands looking down on the action, you would need binoculars to see the width of that line. And the most important point—how much of that last inch is contributed by human activities? The equivalent of a line as thin as a quarter standing on edge.[19]

And how much has CO_2 increased in the atmosphere over the past 160 years? Approximately 35 percent. That's a one-third increase in a century and a half; and the increase is clearly within historical norms. Paleo-climate researchers are quick to reveal data illustrating that in the Eocene period (50 million years ago) CO_2 was likely six times higher than today. In the Cretaceous period (90 million years ago) it was as much as seven times higher than today, and in the appropriately named Carboniferous period (340 million years ago), carbon dioxide is thought to be nearly twelve times that of current levels.[20] Many surmise that dinosaurs were able to grow to such monstrous sizes because of the indescribable abundance of foliage fed by the heightened levels of CO_2 present during their era.

Brilliant research conducted by Michigan State University professor emeritus of horticulture Sylvan H. Wittwer indicates that with a tripling of CO_2, roses, carnations, and chrysanthemums experience earlier maturity and have longer stems and larger, longer-lasting, more colorful flowers, with yields increasing up to 15 percent.[21] Rice, wheat, barley, oats, and rye show yield increases ranging to 64 percent. Potatoes and sweet potatoes produce as much as 75 percent more. Peas, beans, and soybeans show increases to 46 percent. The effects of carbon dioxide on trees, which cover one-third of earth's landmass, may be even more dramatic. According to Michigan State's forestry department, trees have been raised to maturity in months instead of years when the seedlings were raised in a tripled CO_2 environment.[22]

It's also important to realize that, like water, all the carbon dioxide that will ever be is present now, because of the "carbon cycle." So, when a major volcano erupts on the Pacific Rim, a lightning-induced forest fire rages in the Rockies, or an Indonesian peat bog eternally smolders, huge amounts of long-stored CO_2 are naturally released into the atmosphere. Carbon dioxide banked in weathering rocks, decaying coral, and decomposing plants is also constantly meandering through the cycle. It's no different with the carbon cached in fossil fuels; upon consumption it's released into the atmosphere, where it's temporarily held and finally absorbed by a variety of repositories, or "sinks."

As for the carbon sinks, the most obvious is the atmosphere itself, where, as best we can determine, the level of carbon dioxide generally fluctuates between .03 and .04 percent. However, the largest sink is the earth's collective bodies of water, which hold immeasurable stores of dissolved CO_2. That submerged carbon dioxide is used to form the hard exoskeletons of creatures such snails, shellfish, and coral. The ocean's floor is also rich in sedimentary limestone, a petrified modification of CO_2, also known as *calcium carbonate* (the same stuff you take to ease a tummy ache or heartburn). Carbon dioxide is *everywhere*. It's abundant in ancient peat moss deposits, coal seams, natural gas and petroleum reserves, as well as newly fallen autumn leaves, recently felled trees, and in human and animal bodies and corpses. Through the process of decay, the carbon stored in all organic substances is eventually released back into the air as inorganic carbon dioxide, to eventually be reworked into the carbon cycle.

CARBON FOOTPRINT

Al Gore loves to use carbon dioxide in a game of sleight of hand that he plays in all of his media presentations. Gore simultaneously shows two graphs: one tracking CO_2 over the last 650,000 years, alongside another tracking temperature over the same period, embellished with the aforementioned hockey stick spike. The casual observer is struck by the way these two graphs seem to mimic each other.

In his film, with a gleam he addresses the camera and says regarding the graphs, "Incidentally, this is the first time anybody outside of a small group of scientists have seen this image."

Gore only allows a brief glance at his graph, making it impossible to dissect 650,000 years of data. However, if allowed a closer examination, the viewer would see that atmospheric concentrations of CO_2 *always* fluctuate and CO_2 is a lagging indicator of temperature change, not a leading indicator.

Atmospheric carbon dioxide is measured in parts per million (ppm), and currently is 380 ppm. CO_2 reached a low of about 180 ppm during several periods of extreme cold over the past 400,000 years, but always rose to over 300 ppm between these frigid periods. According to ice core records, at the end of each of the last three major ice ages, atmospheric temperatures rose 400 to 1,000 years *before* CO_2 levels increased.[23] And, in more recent times, Al's graph reveals that when temperatures began to warm following the LIA, an increase in carbon dioxide *followed.* Likewise, from 1940 to 1970—the temperature dropped while the carbon dioxide levels rose.

Nevertheless, given the historical record of CO_2, the knowledge that it's part of a cycle, and the fact that it's a lagging indicator of change, environmentalists would have you believe that it's a toxin. It's stunning how they have so perfectly pulled this caper off. A perfectly natural atmospheric gas is now a vilified pollutant. But we're told there is a cure: rigid government intervention and mitigation of one's personal carbon footprint.

Your *carbon footprint* is the amount of carbon *you* are personally responsible for emitting. This includes using energy to illuminate, heat, and cool your home; wash your clothes; and enjoy your big-screen TV. It also entails the gasoline you use to drive your car, truck, tractor, snowmobile, or boat. If you have a diet that includes beef, your carbon footprint is enlarged because of the greenhouse gas associated with cow pies and bovine flatulence. Fly for business or pleasure? That drives up your footprint too—same thing with the use of a fireplace or barbecue.

Regarding personal carbon emissions, James Hansen recommends putting "a flat rising fee on carbon. It will affect consumers and gradually change lifestyles. But give the collected money to the public in an honest transparent monthly dividend. People with lavish life styles will pay more in increased energy costs than they get in their dividend...and they will see that their personal decisions make a difference."[24]

Hansen is preaching the green gospel of sustainable development and social equity: penalize the rich with a pollution tax, which will modify their lifestyles. Then give the tax revenue to the poor in the form of a monthly dividend.

Barack Obama agrees with Hansen's plan. As a senator, one of the few pieces of legislation he sponsored was the Global Poverty Act. The act would require the U.S. to achieve specific and measurable goals consistent with the United Nations Millennium Resolution, including "mobilizing and leveraging the participation of businesses and public-private partnerships; [and] coordinating the goal of poverty reduction."[25] The goal was to create a system of fees to collect and give $75 billion a year to the world's poor: The corresponding press release from Senator Obama's office declared, "In 2000, the U.S. joined more than 180 countries at the United Nations Millennium Summit and vowed to reduce global poverty by 2015. We are halfway towards this deadline, and it is time the United States makes it a priority of our foreign policy to meet this goal and help those who are struggling day to day."[26]

Jeffrey Sachs, who runs the associated UN Millennium Project, has confirmed that his plan is to force the U.S. to pay 0.7 percent of gross national product (that would be about $75 billion in addition to the roughly $50 billion the federal government currently allocates overseas). The only way to raise that funding, Sachs affirms, "is through a global tax, preferably on carbon-emitting fossil fuels."[27]

Fortunately Obama's Global Poverty Act never passed.

Interestingly, Achim Steiner, who heads the UN Environmental Program, shared with world leaders at the 2009 G-20 summit in London a report from his office suggesting that donations of just 1 percent of a country's gross domestic product could bankroll a "Global Green New Deal" to the tune of $750 billion annually. He even floated the option of adding a $5 tax for every barrel of oil consumed by wealthier nations, claiming the surcharge "would generate $100 billion per annum."[28]

A U.S. version of Steiner's Green New Deal nearly became law prior to the Republicans taking back the House of Representatives in 2010. Democratic representatives Henry Waxman from California and Ed Markey from Massachusetts sponsored a massive 932-page global warming bill in May, 2009.

Buried in Section 2201 of H.R. 2454 was the "Energy Refund Program for Low Income Consumers." This portion of the bill would qualify those having an income of up to 150 percent of the poverty level (about $30,000) for a yet-to-be-determined "energy refund," to be "provided in monthly installments via direct deposit into the eligible household's designated bank account."

Thankfully, this bill did not pass, but it remains a part of the master plan.

CAP-AND-TRADE

The preferred means of taxing carbon and distributing the proceeds is a program known as *cap-and-trade*. Through this plan, federal bureaucrats will have the power to determine how many tons of CO_2 individual companies may emit annually—that's the "cap." In the crosshairs of the carbon cops are power plants, oil refineries, cement manufacturers, livestock and dairy producers, farmers, manufacturing plants, paint companies, and commercial and residential buildings. As detailed in the Waxman-Markey bill, as well as the "American Power Act" proposed in May 2010 by Democratic senators John Kerry of Massachusetts and Joe Lieberman of Connecticut, when a business entity exceeds its annual carbon allowance, it may purchase additional "carbon credits" from a government-sanctioned exchange—that's the "trade." The exchange would be run much like a commodities market, complete with institutional spot trades, arranging forward and futures contracts, trading carbon derivatives, and allowing for healthy spiffs to be garnered off the action.

Likewise, should a targeted company remain below its annual carbon cap, it may use the exchange to sell its extra credits to another firm that blew through their limit and is in need of a carbon offset. Firms could also save their credits for future use. Of course, in keeping with Washington, D.C., cronyism, the Kerry-Lieberman plan allows Congress to freely dole out credits to certain preferred businesses and sectors.

Think about this sham for a moment. Cap-and-trade means a perfectly natural component of the atmosphere will be traded like corn, wheat, pork bellies, gold, or oil. Bernie Madoff has to be envious. If instituted nationally, cap-and-trade could well become the biggest financial scam in the history of the world. Money will literally be made out of thin air by slick investors working within a government-sanctioned casino. Market manipulators will have a field day—when the supply of carbon credits is high, the price per unit will be low, and vice versa. Besides manipulation, a large corporation with an ample supply of credits on their ledger could easily destroy the bottom line of a competitor in desperate need of offsets.

In the U.S. the designated carbon-trading house was to be the Chicago Climate Exchange (CCX). CCX was owned by Climate Exchange PLC, which also held the European Climate Exchange (ECX), a firm that handles 90 percent of the world's carbon trades. Al Gore's Generation Investment Management (GIM) fund held a 20 percent stake in ECX and a significant slice of CCX. Many reputable publications, *Human Events* among them, wrote about GIM appearing "to have considerable influence over the major carbon-credit trading firms that currently exist... including CCX.[29]

CCX was founded in 2003 by Richard Sandor, a University of California–Berkeley economics professor. Mr. Sandor candidly told the *Wall Street Journal* that CCX's main goal was "to help develop a commodity that has financial value under any possible future US law that...regulates greenhouse-gas emissions." Asked whether a carbon polluter might cut its greenhouse-gas emissions without the financial pressure of cap-and-trade, Sandor responded that the question is "quite interesting, but that's not my business." Sandor continued, "I'm running a for-profit company."

At one point, CCX had more than four hundred members, including Fortune 100 companies and major utilities.

Goldman Sachs owned a large share of ECX, and a 10 percent stake in the Chicago Exchange. Indeed, Goldman's website still boasts:

> Goldman Sachs is active in the markets for carbon emissions... Additionally, we have created new financial products to help our clients manage the risks posed by climate change. In September 2006, we made a minority equity investment in Climate Exchange PLC, which owns several European and US trading platforms that facilitate trading in environmental financial instruments: the European Climate Exchange (ECX), [and] the Chicago Climate Exchange (CCX) and the newly created California Climate Exchange (CaCX)[30]

In addition, the World Bank joined CCX in June 2006, and currently operates a global Carbon Finance Unit that conducts research on how to effectively develop and trade carbon credits.[31]

However, once it became apparent that Waxman-Markey and Kerry-Lieberman had no chance for passage in the near term, CCX became a toxic asset and was scooped up for pennies on the dollar by a firm known as ICE, the Intercontinental Exchange, Inc. ICE is an American firm specializing in Internet-based marketplaces that trade futures, over-the-counter energy and commodity contracts, as well as financial derivative products. Along with CCX, ICE also bought ECX, which was worth $622 million.[32] ICE chairman and CEO Jeff Sprecher said his company wasn't interested in CCX, but instead saw a better business opportunity within the European exchange.[33]

No one is sure exactly what the shareholders of CCX and ECX received in return for the buyout, but one can guess it was substantially less than the owners envisioned. In fact, I believe it's just one more pin-prick that's been causing Gore to burst of late. Eco-socialists believe that in time they will get their cap-and-trade system. But in the meantime, they're thrilled about a social engineering feature that's likely attached to your home, right now.

SMART METERS

I have a habit of reading each page of every energy bill produced by Congress. So when I read the Federal Energy Act of 2005, signed into law by George W. Bush, Section 1252 grabbed my attention. It states, "It is the policy of the United States that...demand response [technology]...shall be encouraged."[34]

Demand response technology is the *smart meter*, a device that has likely replaced the old, spinning electric or gas meter outside your home. The smart meter logs your energy usage, minute by minute, and remotely sends that data back to your utility provider. In other words, the smart meter monitors your home's carbon footprint. The initial stated goal of this technology was to enable your utility company to collect information that will allow them to charge you more for energy you use during peak periods; it's something called *time-based rate schedules*. Again, quoting from the law: "Not later than 18 months after the date of enactment of this paragraph, each electric utility shall offer each of its customer classes... a time-based rate schedule under which the rate charged by the electric utility varies during different time periods"[35]

To prevent a public backlash from occurring once citizens realized how Big Brother was soon to be attached to their home, the law also authorized the secretary of energy to educate "consumers on the availability, advantages, and benefits of advanced metering and communications technologies," and to work "with States, utilities, other energy providers and advanced metering and communications experts to identify and address barriers to the adoption of demand response programs."[36]

The only "advantages" and "benefits" from the smart meter are to those who seek control of your life.

The smart meter is designed to eventually connect to the *smart grid*. The smart grid was first brought to the public's attention during the 2008 presidential campaign, as Obama spoke of the "need to modernize our national utility grid."[37] And, like the smart meters, the smart grid was also covertly introduced in the Energy Act of 2005. Section 1221 permits the construction of "a national interest transmission corridor designated by the [Energy] Secretary."[38]

In the months following the bill's signing, a spark of concern began to develop: new transmission lines associated with this "transmission corridor" were going to be constructed through many states without their approval, violating the constitutional provision of state's rights. Eventually, token Congressional hearings were conducted. One of the more vocal opponents was Bill DeWeese, a Pennsylvania state representative who had learned that fifty-seven of Pennsylvania's sixty counties had been declared a part of the federal corridor.[39] DeWeese spoke before Congress, arguing that the Federal Energy Regulatory Commission was about to "usurp the traditional role of states and their administrative agencies to review and approve the location and construction of high voltage transmission lines."[40]

DeWeese was correct—states were shut out of the discussions.

It wasn't until 2007 that the mysterious federal corridor was finally christened in the Energy Independence Security Act. Buried in its 310 pages, an entire section was devoted to the grid. Some three thousand miles of electric lines would be strung, virtually updating the entire nation's existing infrastructure. However, proving that most members of Congress (and perhaps even the president) didn't know what they had agreed to—probably because they never read the bill, or, if they did, didn't bother to ask any questions—on December 20, the day *after* the legislation became law, a report from the Congressional Research Service was distributed to members of Congress, describing how the Smart Grid works:

> The term Smart Grid refers to a distribution system that allows for flow of information from a customer's meter in two directions: both inside the house to thermostats and appliances and other devices, and back to the utility. It is expected that grid reliability will increase as additional information from the distribution system is available to utility operators. This will allow for better planning and operations during peak demand. For example, new technologies such as a Programmable Communicating Thermostat (PCT) could connect with a customer's meter through a Home Area Network allowing the utility to change the settings on the thermostat based on load or other factors. PCTs are not commercially available, but are expected to be available within a year.[41]

The *programmable communicating thermostat* is a product that your utility company is, no doubt, pushing—in fact, you may already have one. It's a device attached to your home thermostat that can be remotely controlled by your utility company. So, if there's an energy shortage and you're running your air-conditioning unit, your utility company could remotely alter the temperature in your home, or just turn the AC off. The *home area network* (HAN) is still in development and will interface with all government-approved Energy Star appliances and other electrical devices within your home, including your TV, computer, and lights.

And then there's the grid. Unlike the current electrical grid crisscrossing the nation, the smart grid will be wired for broadband; however, few politicians will tell you why. In an address to Northern Michigan University in 2011, President Obama cleverly sold it like this: "We want to invest in the next-generation of high-speed wireless coverage for 98 percent of Americans. This isn't just about a faster Internet or being able to friend someone on Facebook. It's about connecting every corner of America to the digital age."

That "digital age" Obama spoke of is the age of Big Brother monitoring your carbon footprint.

The smart grid's interactive broadband capability will enable your home's PCT, HAN, and smart meter to be connected and communicating with your utility provider. Once complete, the utility company will be your government-sponsored Big Brother, constantly monitoring and regulating your carbon footprint. With a bureaucratic keystroke, any electrical device in your home could be selectively turned off—or on—without your approval.

Originally, according to Section 1306 of the Energy Independence Security Act, funding the smart grid was to be shared by the various states. However, this funding mechanism posed a serious problem. Most states were working with rapidly thinning budgets, and it was unlikely that they would be willing to invest the billions necessary to upgrade to the new grid. Likewise, most utility companies were unable to invest the necessary capital for improvements. The dilemma was rectified in the 2009 stimulus bill, which authorized $16.8 billion in direct spending by the U.S. Department of Energy's Office of Energy Efficiency and another $4.5 billion to upgrade the nation's grid, plus at least $2.8 billion for installing broadband—another component critical to the success of the smart grid. Thus, with one of the first strokes of his golden presidential pen, Obama lifted the burden of funding the Orwellian smart grid off the slumping shoulders of the states, and onto the backs of the American taxpayers.

Liberty-seeking Americans understand the smart meter and smart grid for what they are: a giant step toward tyranny.

SIX

GREEN GOSPEL

RELIGION IS POWERFUL MEDICINE.

History is loaded with examples of how people linked by a common, deeply held ideology are capable of accomplishing the wonderful—or carrying out the unthinkable—in the name of their convictions.

Slavery in the Western world, for example, was brought to an end in the 1800s because of the tireless campaigning of evangelical Christians such as William Wilberforce, Hannah More, and John Newton. Contemporary natural disasters, such as floods and earthquakes, are always followed by significant acts of charity from Bible-based organizations, including the Salvation Army and Franklin Graham's Samaritan's Purse. The individuals involved in such works are zealous for doing good because of what they have read, studied, and absolutely profess.

But such piety can also produce evil—look no farther than the present Muslim world, which is churning out countless thousands who memorize pages of the Koran; subject themselves to fiery, hate-packed sermons; and too often respond with a passion to destroy the nonbeliever—even taking their own lives in the process—in the name of their beliefs.

However, traditional religion is ceding ground to a doctrine that is rapidly gaining new adherents. Like its conventional cousins, this contemporary creed inspires, transforms, and actuates its adherents. This fresh dogma is being preached in America's public schools and elite private colleges, as well as in our churches and synagogues. It's a pietism that has successfully mixed junk science, humanism, and paganism into an earth-worshipping stew.

It's a spirituality known as "biocentrism," and its deity is "Gaia." Let's begin with Gaia—and no, it's not a PR stunt hatched by Lady Gaga. We could laugh about that. The reality of Gaia-worshippers is no laughing matter.

Gaia was the mythological Greek goddess of Earth—Mother Earth, if you will. She was resurrected from the ash heap of the ridiculous by an eccentric scientist named James Lovelock, along with Lynn Margulis, the first wife of astronomer Carl Sagan, in the 1970s. Lovelock has written a series of best sellers on the subject, but his first was *Gaia: A New Look at Life on Earth*, published in 1982. Lovelock's Gaia hypothesis contends that the earth is a singular living organism that has the capacity to regulate, repair, and heal itself under natural conditions. However, according to Lovelock, Gaia is suffering from a severe malady—*people*. He believes that human population has grown to the point that it is overwhelming Gaia's capacity to restore herself, and therefore, she is doomed to destruction unless the human species stops the onslaught of pollution and the use of Gaia's natural resources.

In other words, like Paul Ehrlich, Lovelock sees the human race as a disease.

Initially, Lovelock's work was soundly rejected by the scientific community as being from Kooksville, but then (as you will soon see) the United Nations stepped in—with substantial support from Maurice Strong, by the way. Soon, other scientists rallied to try to provide evidence to Lovelock's claims.

GAIA IS THE GOD; BIOCENTRISM, THE RELIGION

Like Marx and his disciples, Lovelock and Margulis are materialists, and materialists have always wrestled with one monumental dilemma: how did life originate? More specifically, how did something come from nothing? Even if there was a big bang, where did the material originate to make it happen? How did the stuff that was produced by the bang become alive?

And regarding the most advanced species to evolve from the primordial goo—Homosapiens—there is another monumental question: how does one explain *consciousness*?

The doctrine of biocentrism wildly strives to supply the answers.

"Bio" is Greek for "life." "Centrism" also stems from the Greek word *kentron*, which means "at the center." Biocentrism believes that all forms of life are equally valuable, and humans are not supreme beings to be valued more highly than all other species. Biocentrism provides a new label for what Karl Marx and Frederic Engels believed.

Contrast biocentrism with the viewpoint that has been held without question until minds like Marx's came along: *anthropocentrism*, which declares that human beings and human society are the central focus of existence.

However, the biocentrist takes yet another leap and is convinced that because humans are the more complex species, they have the greatest ability to harm all others. Hence, biocentrists have adopted a prejudice against humans. It all flows back to Marx's laws of matter and how humans require strict regulation lest they kill one another and destroy the planet.

Biocentrism's chief pitchman is Dr. Robert Lanza, the technician at the forefront of the highly controversial embryonic stem cell research. Such experimentation requires cloning and destroying a human embryo in order to harvest key components within the embryo to develop cures and therapies for humankind's physical ills. But despite billions of dollars in research, embryonic stem cell research has failed miserably, as not a single cure or therapy has been developed from such morbid experimentation. Adult stem cell research (culled from human tissue via a simple biopsy), on the other hand, has yielded spectacularly successful cures and therapies for a host of ailments—without the destruction of human embryos.

Lanza's biggest sale of biocentrism has focused on trying to explain consciousness. Says Lanza, "We don't begin to understand where the Big Bang came from even if we continually tinker with the details. Indeed, every theorist realizes in his bones that you can never get something from nothing, and that the Big Bang is no explanation at all for the origins of everything."[1]

Lanza encourages his devotees to scrap considering the origins of life and instead allow themselves to undergo a paradigm shift whereby they come to properly perceive that humans are not the most significant beings on the planet, nor the most notable beings of all time. According to him, "Without perception, there is, in effect, no reality. Nothing has existence unless you, I, or some living creature perceives it, and how it is perceived further influences that reality. Even time itself is not exempted from biocentrism."[2]

Not only does the biocentrist believe life created itself and that the human brain has essentially tricked us into accepting a perceived reality that may not exist at all, but biocentrism also believes that all forms of life are intrinsically connected. In other words, every activity in which we're engaged has a distinct global impact. Thus, biocentrism melds well into the Gaia model of the earth being a single living organism that is now being harmed by the activities of all humans.

"One member instantly influences the behavior of the other—even if they are separated by enormous distances," says Lanza. "They are intimately linked in a manner suggesting there's no space between them, and no time influencing their behavior."[3]

An environmental scientist and fellow biocentrist who is received like a rock star on college campuses nationwide is Paul Taylor. He provides further detail to the biocentrist position, arguing that each living component on earth is what he calls a "teleological-center-of-life" that possesses a well-being of its own and "inherent worth." These teleological-centers-of-life, which include everything from a single-cell plankton to a redwood tree to a mountain lion to a polar bear, can be enhanced or harmed by human activity and are therefore entitled to moral respect.[4] Furthermore, Taylor preaches that the intrinsic value of wild living things generates a prima facie moral duty on the part of humans to preserve and promote the well-being of these forms of life, and any practice that treats those "beings" with a lack of respect is intrinsically wrong.[5]

Biocentrism is the new religion—it can't be proved, but it is believed. Add Gaia as the deity of this new faith and you've got some powerful mojo.

Thus, the actions of the Gaia-inspired biocentric environmentalists are not a lot different from those of God-fearing pro-lifers who believe "it's a baby, not a choice," and will go to great lengths to stop abortions. The biocentrist perceives an ancient tree as a great-great grandfather, the red-legged frog as a cousin, and that weed you want to eradicate from your lawn, as a friend; whatever has to be done to defend these life forms must be done.

UN PROMOTES THE GREEN RELIGION

Knowing that the formal rollout of the green agenda would be difficult for conservatives in the United States to accept (especially given that most American conservatives are respectful of their Judeo-Christian heritage), in the late eighties, a decision was made by the UN to begin influencing the primary pillar of traditional thought in America—the church. Lovelock was recruited as the primary evangelist.

In 1988, the UN Global Committee of Parliamentarians on Population and Development (a group dedicated to seeing the planet's population reduced) held a forum in Oxford, England, with Lovelock as the featured speaker. "On Earth," Lovelock preached, "she [Gaia] is the source of life everlasting and is alive now; she gave birth to humankind and we are a part of her."[6] He also professed with great hyperbole that global warming is the result of a violent human assault upon the earth: "She may be unable to relax because we have been busy removing her skin and using it as farm land, especially the trees and forests of the humid tropics...we are also adding a vast blanket of greenhouse gases to the already feverish patient."[7]

Gaia's introduction to the UN was so well received that a second forum was hosted in 1990 by the Soviet Union's communist chief, Mikhail Gorbachev. Many of the world's prominent religious leaders attended the Moscow bash, which featured Carl Sagan's appeal for science and religion to "join hands" in a new ecological alliance—quite an overture for an outspoken atheist. More than a hundred religious leaders would eventually endorse his entreaty, including author and Holocaust survivor Elie Wiesel; Ahmed Kuftaro, noted as Islam's ambassador to the world; Joseph Cardinal Bernadin, archbishop of Chicago; Rev. Theodore Hesburgh, president emeritus of Notre Dame University; Native American spiritual leader Oren Lyons; Rev. Ronald Thiemann, dean of the Harvard Divinity School; and Rev. Dr. Milton B. Efthimiou of the Greek Orthodox Archdiocese of North and South America.

Seizing the moment, Gorbachev took to the microphone and spouted, "Perestroika [the restructuring of the Soviet economic and political system] has changed our view of ecology; only through international efforts can we avert tragedy." He called for each nation to begin crafting a wish list of items to be included in the official agenda for the twenty-first century—Agenda 21—to be drafted at the forthcoming 1992 UN Earth Summit in Rio de Janeiro. Gorbachev also made an additional plea for aid in the formation of his personal dream of the "Green Cross," an international emergency task force that could supposedly be rushed to the scene of any ecological disaster.[8] James Hansen of NASA was also present at the event and vigorously supported the appeal.[9]

By the way, Gorbachev's dream of a Green Cross came to fruition in 1993. The International Green Cross currently boasts a cast of outspoken environmentalists on their board, including actor Robert Redford and media mogul Ted Turner; Gorbachev remains the organization's founding president. As one might have predicted, the Green Cross never did launch an environmental strike team to clean up pollution across the globe. Instead, the group has been, and continues to be, committed to eliminating capitalism and liberty from the planet, as it seeks, in accordance with its mission statement, to "ensure basic changes in the values, actions, and attitudes of government, the private sector, and civil society necessary to build a sustainable global community."[10]

In other words, Gorby hasn't changed his stripes one iota. He's still a communist fighting the cold war. This time, though, instead of employing Soviet spies, submarines, and a nuclear arsenal, he's using the high-profile environmentalist celebrities and some spiritual-sounding lingo in his effort to emasculate America.

Appearing on Charlie Rose's PBS television program, Gorbachev said, "We are part of the cosmos...cosmos is my god. Nature is my god...I believe that the twenty-first century will be the century of the environment, the century when all of us will have to find an answer to how to harmonize relations between man and the rest of Nature...we are part of Nature."[11]

GREEN RELIGION COMES TO AMERICA

A year after the Moscow conference, another meeting of religious leaders and scientists was assembled in Washington, D.C., this time with members of Congress invited. The event was sponsored by the newly formed

National Religious Partnership for the Environment (NRPE). During the conference, a declaration was drafted that read: "We believe a consensus now exists, at the highest level of leadership across a significant spectrum of religious traditions, that the cause of environmental integrity and justice must occupy a position of utmost priority for people of faith."[12]

Following the event, eleven executive officers of major national environmental groups sent a group letter endorsing the declaration, especially noting the desire of the NRPE and its supporters to address the "struggles for environmental justice by poor, minority and indigenous peoples."[13] Among the groups were the National Audubon Society, the Natural Resources Defense Council, the Sierra Club, the Environmental Defense Fund, and the World Resources Institute.

Two years later, a luncheon was held to honor the NRPE's work. As hoped for, the organization's cunning efforts were now drawing in a broader spectrum of more conservative religious elements, and this occasion included representatives from the National Baptist Convention, the National Association of Evangelicals, World Vision, and Intervarsity Christian Fellowship. Vice President Al Gore was also present and manned a post-luncheon press conference in which he told the media that the NRPE's activities "will trigger the beginning of grassroots activity in tens of thousands of religious congregations across the country."[14]

The environmental movement was successfully co-opting mainstream religion. Witnessing the proceeding was Carl Sagan who said, "Separately, neither science nor religion could solve the problem of redeeming the environment from the shortsightedness of the last few decades."[15]

Having been established as legitimate, the NRPE next successfully raised and spent millions to spread its gospel of green to fifty-three thousand congregations in America, including every Catholic parish, plus every Reform and Conservative synagogue in the nation. The sent materials were ongoing and provided information for clergy and lay leadership training on environmental topics and issuing legislative updates and action alerts. A "1-800-Green Congregation Hotline" was also established to help organize grassroots religious environmental activities, as well as a 171-page who's who directory of cooperating environmental organization partners.

Illustrating the influence the NRPC was having on the Christian community, in 1995–1996 the organization held several exclusive conferences for private Christian colleges and universities, teaching the schools how to effectively incorporate environmentalism within the context of Christian stewardship.[16] The plan was incredibly productive. Today even the staunchest born-again Christian colleges are espousing an apocalyptic message regarding the state of the planet. Wheaton College in Illinois, the school that taught arguably the most prolific evangelist in history, Billy Graham, has a brochure promoting their 2011 Environmental Studies Program, which sounds as if it could have been written by James Lovelock himself: "There is growing evidence that humankind has had such a negative impact on God's creation that unpredictable changes will jeopardize the future of life on earth."[17]

Can anyone who ever saw a Billy Graham crusade envision him incorporating such a theme into one of his sermons?

Imagine the piano playing "Just as I Am," Graham's patent theme song for the wayward as they were invited to the altar for a time of reflection and repentance. Then Reverend Graham steps to the microphone and pleads, "You need the Lord, my friend—but to really get right with God, you need to reuse, recycle, drive less, and go solar."

GREEN CITIZENS

In his book *Earth in the Balance*, Al Gore plays the role of a fire-and-brimstone preacher from the First Green Church as he articulates how to avoid a "wrenching transformation of society":

> I have come to believe that we must take bold and unequivocal action: we must make the rescue of the environment the central organizing principle for civilization. Adopting a central organizing principle—one agreed to voluntarily—means embarking on an all-out effort to use every policy and program, every law and institution, every treaty and alliance, every tactic and strategy, every plan and course of action—to use, in short, every means to halt the destruction of the environment and to preserve and nurture our ecological system. Minor shifts in policy, marginal adjustments in ongoing programs, moderate improvements in laws and regulations, rhetoric offered in lieu of genuine change—these are all forms of appeasement, designed to satisfy the public's desire to believe that sacrifice, struggle, and a wrenching transformation of society will not be necessary.[18]

Gore is playing a devious game here. On the one hand, he claims the organizing principles to "rescue of the environment" should be "agreed to voluntarily." However, in the very next breath, he advocates "an all-out effort" to use every known lever of government—policies, programs, laws, institutions, treaties and strategies—to halt "the destruction of the environment," a.k.a. the end of life on earth. Gore is using blatant scare tactics to coerce his audience into modifying their behavior. It's not much different from the religious leader who threatens you with hell and damnation unless you pony up when the plate is passed. The goal of Gore's "all-out effort" to supposedly save the planet is to force upon us behavior modification. Policies to limit automobile choices are employed. Programs to subsidize the proliferation and use of alternative energy sources are practiced. Laws banning incandescent lightbulbs are passed. Treaties to reduce carbon dioxide are agreed to. Strategies to institute complex cap-and-trade schemes are devised. And to young minds in our nation's schools, action and subsequent behavior is taught to be the norm. It's a brainwashing—or green washing—that's being championed from the highest levels of government.

"Today, I promise you," stated education secretary Arne Duncan at an invitation-only gathering for three hundred educational leaders, "that under my leadership, the Department of Education will be a committed partner in the national effort to build a more environmentally literate and responsible society."[19]

The amount of money we spend on education per student in the U.S. is the highest in the world, and yet from the most recent internationally recognized study (the Program for International Student Assessment) we know that our students are merely average in reading (ranked fourteen out of the thirty-three countries sampled), awful in math (twenty-third), and subpar in science (seventeenth). How will teaching environmental literacy improve those vital statistics?

Duncan continued, "Right now, in the second decade of the twenty-first century, preparing our children to be good environmental citizens is some of the most important work any of us can do. It's work that will serve future generations—and quite literally sustain our world."

And how does this administration propose to do it? According to Duncan, through new federally subsidized school programs that, beginning as early as kindergarten, teach children about climate change and prepare them "to contribute to the workforce through green jobs."

At the end of the day, the goal of education in the United States is about "educating the next generation of green citizens."

And in training these "green citizens," Duncan says, teachers must employ the doctrine of biocentrism to make sure students know that their poor choices now will endanger the next generation: "Educators have a central role in this. A well-educated citizen knows that we must not act in this generation, in ways that endanger the next. They teach students about how the climate is changing. They explain the science behind climate change and how we can change our daily practices to help save the planet."

To better accomplish this task, Duncan explained to the group that the Department of Education had prepared an initiative called the "Blueprint for Reform" that will "transform our economy, to protect our security, and save our planet."

President Obama actually proposed $265 million for this program in his fiscal 2011 budget. Fortunately the plan was never approved by Congress. We must make sure that continues to be the case, because Duncan let the cat out of the bag when he went on to say that the money for the Blueprint for Reform would be used to "build the science of sustainability in the curriculum, starting in kindergarten and extending until the students graduate high school."

ENVIRONMENTAL CATECHISM

Actually, even without Obama's Blueprint, our kids are already getting their brains tinted green through indoctrination processes conducted by their teachers. Take, for example, America's oldest school—Boston Latin.

Boston Latin is a public high school, founded in 1635, one year before Harvard College was established, that boasts a patriotic list of former pupils including Benjamin Franklin, Samuel Adams, and John Hancock. Today, Boston Latin also boasts being the birthplace of the Youth Climate Action Network (Youth CAN). Youth CAN has now expanded nationwide with an uncompromising vision: "Imagine many youth climate action groups speaking with one voice, insisting that legislators make the necessary changes pertaining to global warming."[20] Youth CAN's moblike tactics include organizing protests, boycotts, and civil disobedience in the form of hijacking class time with teach-ins.

After a screening of Gore's *An Inconvenient Truth* in 2007, Youth CAN started brainstorming ideas for incorporating an avant-garde sustainability curriculum into the entire school's list of offerings. As one club member told the *Boston Globe*, "I think schools can do a much better job integrating sustainability into the curriculum and not just have one unit about climate change but incorporate entire themes."[21]

And the students are getting their way.

Headmaster Lynne Mooney Teta and faculty moved forward with the sustainability curriculum, and in the fall semester of 2010, students began being exposed to sustainability issues in a wide range of courses crossing all disciplines. A physics teacher said he "envisions sustainability issues as the 'backbone of the curriculum,' integrated in many subject areas."

A peek at the school's curriculum guide illustrates how the students at Boston Latin are actually being frightened into submission. A summary of their environmental science course states:

> At the beginning of a new millennium, civilization has reached crossroads. Unable to halt the technologies that continue to exploit Earth's resources, and facing the prospect of a continuing rise in the human population, we find ourselves in a situation where an enormous human-environment experiment is being enacted. Environmental science stands at the interface between humans and the Earth: it explores the interactions and relations between them. The issues to be explored include global warming, species extinction, air pollution, toxic wastes, overpopulation, recycling, water, waste removal, and biodiversity.[22]

According to the *Globe*, the curriculum alterations were carried out with the assistance of the Children's Environmental Literacy Foundation (CELF). The CELF website brags of its green-washing capability:

> Sustainability education must begin earlier in students' academic life. We focus on K–12 as formative years for shaping thinking, attitudes, values and behaviors... Once students grasp the connections between a stable economy, a healthy environment and equitable social systems, and their role as global citizens, they are successfully launched into an already more sustainable world.[23]

By forcing this powerful green catechism on youth in their formative years, CELF is not only fostering the ruin of our basic educational system in the U.S., but is abetting a mind-set that is running on a fear-based platform.

Take the case of nine-year-old Rafael de la Torre Batker in Tacoma, Washington. Rafael was riding in his parents' car and, according to the *New York Times*, "was worried about whether it would be bad for the planet if he got a new set of Legos."[24]

Rafael's class had just watched *The Story of Stuff*, an animated anti-capitalist diatribe by former Greenpeace employee Annie Leonard. The program is a big hit among teachers looking to green up their schools' environmental curricula. Leonard claims her video has been viewed by more than three million people online, and some seven thousand copies of the DVD have been sold. The Environmental Protection Agency, in conjunction with the Corporation for Public Broadcasting, is now paying Leonard to produce additional environmental propaganda for distribution online, on DVD, and on the air.

Leonard describes herself as an "unapologetic activist," and in the style of a committed biocentrist, isn't shy about painting hyperbolic doomsday scenarios in which evil corporations and selfish consumers end up destroying life as we know it. In Leonard's twenty-minute *Story of Stuff* film, she claims that the production of consumer goods is destroying the planet. Take, for example, little Rafael's Legos, which are made of plastic, a by-product of oil. Oil is a natural resource that is extracted from the earth. Leonard makes kids believe this is wrong and that Lego is guilty of committing an immoral act. Indeed, corporate America is characterized in the film as a fat man sporting an old-fashioned top hat with a big dollar sign emblazoned above the brim.

"Extraction which is a fancy word for natural resource exploitation, which is a fancy word for trashing the planet," Leonard told the *Times*.

Leonard also had a popular book being used in the classroom by thousands of teachers: *The Story of Stuff: How Our Obsession with Stuff Is Trashing the Planet, Our Communities, and Our Health—and a Vision for Change.*

"There's no way around it," Leonard says. "Capitalism, as it currently functions, is just not sustainable."

And, in keeping with the now well-developed plan to co-opt the church, Leonard has partnered with a group known as Green Faith to produce a lesson plan, designed to be used in private Christian schools and church Bible studies, that emphatically presents a new morality: "When our waste damages Creation, we may not feel aware of the ramifications of our pollution. But God doesn't easily excuse our lack of attention to these concerns. Instead, God sees our carelessness as the precursor to earth's devastation. Our quick forgetfulness is the first step in earth's devastation."[25]

Green Faith is not the only organization pushing an ideology of fear, guilt, and self-loathing; so is the government.

GOVERNMENT-SPONSORED SCAREMONGERING

After the shooting of Arizona Congressional representative Gabrielle Giffords, there was a call to tone down the violent political rhetoric, which Democratic members of Congress claimed was originating from the right. Democratic National Committee chairperson and Florida representative Debbie Wasserman Schultz took to the cameras, stating, "Words matter." She then reminded her fellow public servants that "in terms of civility and tone, we have to set an example."[26]

It's too bad the congresswoman won't have the same conversation with her friends in the environmental community.

Let's start at the top with a recent outlandish quote from Al Gore.

> Global warming, along with the cutting and burning of forests and other critical habitats, is causing the loss of living species at a level comparable to the extinction event that wiped out the dinosaurs 65 million years ago. That event was believed to have been caused by a giant asteroid. This time it is not an asteroid colliding with the Earth and wreaking havoc: it is us.[27]

If such an asteroid did slam into the earth, it's thought that the impact would have been thousands of times more powerful than the most powerful nuclear bomb. The grand Nobel laureate is threatening the public with a catastrophe that defies the imagination and expects to get a free pass. This is dangerous talk that could cause an unbalanced mind to go wobbly.

Let's go next to Al's friend, NASA director James Hansen.

"The climate is nearing tipping points," he said in a 2009 opinion piece published in one of London's most popular newspapers. "Changes are beginning to appear and there is a potential for explosive changes, effects that would be irreversible, if we do not rapidly slow fossil-fuel emissions over the next few decades."

Hansen next describes the apocalyptic warning signs: "As species are exterminated by shifting climate zones, ecosystems can collapse, destroying more species."

Hansen then reveals the demon behind such environmental evil—coal: "Coal is not only the largest fossil fuel reservoir of carbon dioxide, it is the dirtiest fuel. Coal is polluting the world's oceans and streams with mercury, arsenic and other dangerous chemicals... The trains carrying coal to power plants are death trains. Coal-fired power plants are factories of death.[28]

Folks, this is a director of NASA—the agency that (until recently) put men and women in outer space. An irrational mind might just take this maniacal rhetoric to heart, and strap him- or herself to the train tracks in an attempt to halt such a "death train."

Hansen is notorious for his reckless, inciting suggestions. In an interview with a counterculture radio show in San Francisco, he said:

> I tell young people that they had better start to act up. Because they are the ones that will suffer the most. Many of the changes will take time, but we're setting them in motion now. We're leaving a situation for our children and grandchildren which is not of their making, but they're going to suffer because of it. So I think they should start to act up and put some pressure on their elders, and on legislatures, and begin to get some action.[29]

For the record, Dr. Robert Jastrow, a founder of NASA's Goddard Institute of Space Studies, which Hansen heads, reportedly told a trusted friend toward the end of his life that his "one true" regret as a professional researcher was handpicking Jim Hansen to be his successor. Jastrow's friend was Dr. Willie Soon, an acclaimed Harvard-Smithsonian astrophysicist.[30]

NASA isn't the only federal agency associated with climate scaremongering; the Environmental Protection Agency is also in on the game. Posted on the EPA's website is a list of frequently asked questions on global warming. In one response the agency declares, "Climate change health effects are especially serious for the very young, very old, or for those with heart and respiratory problems."[31] Another EPA document states, "Climate change will likely increase the number of people suffering from illness and injury due to floods, storms, droughts, and fires, as well as allergies and infectious diseases."[32]

OBAMA'S GREEN CZAR

And then there is Van Jones, truly a rock star in the environmental movement. Jones was handpicked to join the Obama administration in March 2009 as special adviser for green jobs, or, unofficially, the Green Jobs Czar. No sooner had Jones come on board than conservative media types like me cried foul—the guy was a hip-hop communist in a silk suit.

Speaking to the San Francisco newspaper *East Bay Express*, Jones said he first became radicalized in the wake of the 1992 Rodney King riots, during which time, he was arrested. "I was a rowdy nationalist on April 28th, and then the verdicts came down on April 29th. By August, I was a communist." Adding further details of his commie conversion, Jones explained, "I met all these young radical people of color—I mean really radical: communists and anarchists. And it was, like, 'This is what I need to be a part of.' I spent the next 10 years of my life working with a lot of those people I met in jail, trying to be a revolutionary."[33]

By early September 2009, six months into his Green Czar role, Jones was forced to issue two public apologies, one for a 2004 petition he had signed from the group 911Truth.org that questioned whether Bush administration officials "may indeed have deliberately allowed 9/11 to happen, perhaps as a pretext for war," and the other for remarks uttered in February 2009 during an energy lecture in Berkeley, California, after a woman in the audience asked him why President Obama and congressional Democrats were having trouble moving legislation—even though Republicans, with a smaller majority, didn't have as much trouble earlier in the Bush administration.

"Well, the answer to that is, they're assholes," Jones said, to uproarious laughter. "That's a technical, political science term."

His one-time involvement with the Bay Area radical group Standing

Together to Organize a Revolutionary Movement (STORM), which had distinct Marxist roots, had also become an issue. And then we discovered Jones's advocacy (he is a lawyer) on behalf of death-row inmate Mumia Abu-Jamal, who was convicted of shooting a Philadelphia police officer point-blank in 1981, threatened to develop into a fresh point of controversy.

Jones is another climate change scaremonger.

Shortly before his White House appointment (which was never confirmed by the Senate), Jones was a keynote speaker at the Power Shift 2009 Conference in Washington, D.C., an environmental conference attended by some twelve thousand people. At the height of his address, with the crowd whipped into a wild frenzy, Jones shouted, "This movement is deeper than a solar panel! Deeper than a solar panel! Don't stop there! Don't stop there! We're gonna change the whole system! We're gonna change the whole thing. We're not gonna put a new battery in a broken system. We want a new system. We want a new system!"

The "new system" Jones endorsed is one fashioned by Marx. Fortunately appropriate pressure was placed on the White House to send Jones packing. However, President Obama did not fire Van Jones; instead Jones resigned in September 2009. His resignation lifted him to the status of a martyr amongst the eco-faithful.

ECO-TERROR

We are witnessing a generation being filled with green religion, a faux spirituality that is as dogmatic, and potentially dangerous, as any on the planet. So when an undiscerning individual hears teachers, church leaders, scientists, political representatives and peers uttering ominous predictions about the state of the earth, is it any wonder that some go off the rails and make really bad choices based on the alarming rhetoric they've been force-fed?

For example, I'm sure thousands who have heard the green gospel have decided their best sacrifice for Mother Earth is to spend money they don't have to purchase a hybrid vehicle. Making the decision seems all the easier when a federal or state tax rebate is offered for making the decision to worship at the green altar. For others, the calling is to go vegan. Farm animals and livestock create loads of greenhouse gases. And for the exceptionally devout, the real show of dedication is to have an abortion to fight global warming. A story in the *London Sunday Times* quoted the British government's "green advisor," Jonathon Porritt, who says couples raising more than two children are being "irresponsible" by creating an unbearable burden on the environment.

Porritt, who chairs the British Sustainable Development Commission, says that curbing population growth through contraception and abortion must be at the heart of policies to fight global warming: "I am unapologetic about asking people to connect up their own responsibility for their total environmental footprint and how they decide to procreate and how many children they think are appropriate."[34] He also says that political leaders and green campaigners should stop dodging the issue of environmental harm caused by an expanding population.

I conducted some "man on the street" interviews at an Earth Day festival in far-left-of-center Santa Cruz, California, and had a spokesman for a population-control organization inform me, "Smaller families live better."[35]

And besides those who believe taking the unborn is just for the sake of the environment, there are still others who have gone even farther.

In 2010 James Jay Lee executed a dangerous hostage plot inside the headquarters of the Discovery Channel. Armed with what appeared to be pipe bombs and a cheap pistol, Lee claimed to have been "awakened" by Al Gore's film, *An Inconvenient Truth*. Lee regarded humans as the "most destructive, filthy, pollutive creatures around." His desire was to force the Discovery Channel to fill its programming schedule with "solutions to save the planet." Before he was able to harm innocent life, Lee was shot and killed by police.

Unfortunately, Lee is not the first eco-freak to go off the deep end. In 2005, four years after 9/11, the FBI declared domestic ecoterrorism to be America's number one threat.

Other recent examples of enviro-violence include two large commercial radio towers toppled in Washington state. The site was tagged with the letters "ELF." It seemed to be the obvious work of the Earth Liberation Front. In fact, photos of the destruction were posted on the group's website. ELF has a history of such attacks. In April 2010, ELF member Stephen Murphy was sentenced to five years in prison after admitting he conspired to burn down a condominium development in Pasadena, California.

Daniel Andreas San Diego is on the FBI's Most Wanted list for his alleged involvement in two bombings of biotech firms in the San Francisco area; one of the bombs was loaded with nails. According to the FBI, San Diego bears freaky tattoos resembling burning hillsides on his chest with the words "It only takes a spark," as well as burning buildings on his abdomen and a leafless tree rising from a road on his back.

Of course, the most notorious ecoterrorist is Ted Kaczynski—the Unabomber. Over a seventeen-year period during the eighties and nineties, Kaczynski sent out mail bombs, killing three people and wounding twenty-two. He also managed to sneak a bomb onto a 747 passenger jet flying from Chicago to Washington, D.C. Fortunately, the bomb didn't go off as planned. Kaczynski's reign ended in 1996, shortly after he made public his now-infamous manifesto written in his tiny cabin located in the backwoods of western Montana. In it he opined, "One of the effects of the intrusion of industrial society has been that over much of the world traditional controls on population have been thrown out of balance. Hence, the population explosion, with all that it implies... No one knows what will happen as a result of ozone depletion, the greenhouse effect and other environmental problems that cannot yet be foreseen."

And discovered by the FBI in the Unabomber's hovel? A well-worn copy of Al Gore's *Earth in the Balance*. Kaczynski apparently was quite taken by Gore's missive. The Unabomber's copy was dog-eared, underlined, marked, and tattered.

The question is, why should anyone be surprised when environmental nut jobs mentally detonate? If one takes the warnings of Gore and comrades as truth, planet Earth is a ticking time bomb, and if nothing is done to stop it, we'll all perish. For some, such fearmongering becomes a clarion call to hyperaggressive, unapologetic action.

The infectious perspective of the environmental movement has slithered into every aspect of American life, including our schools, churches and synagogues, and public policy. An entire generation and more have now been raised in a perpetual pall that declares the earth is doomed because of mankind's pollution. These same citizens have been duped into believing that America's experiment with capitalism and free markets has been a complete failure, and the major evidence is climate change. Having been fed a continual diet of junk science and raised with strict environmental regulations, they believe such stratagems to be the norm, and, consequently, they simply accept the untruths without question.

But let's be clear about this movement—it's being driven by a devout communistic and socialistic ideology. Each time you hear an eco-activist or representative of an environmental organization speak, know this: he or she is knowingly pushing the message of Marx and deviously hoping to see the United States changed. These activists will spin, bewilder, and lie. Their ultimate goal is to expunge the most precious profession ever put forth by our Founders: "We hold these truths to be self-evident, that all men are created equal, that they are endowed by their Creator with certain unalienable Rights, that among these are Life, Liberty and the pursuit of Happiness. That to secure these rights, Governments are instituted among Men, deriving their just powers from the consent of the governed..."

And what do the radicals propose in place of the American way?

Tyranny.

Like Marx, they are convinced that we the people left to our own devices, will destroy the planet. Also like Marx, they believe that our inherent desire to improve our lives and raise our standard of living is obscene, and that profit is corrupt. They further conclude that private property is perverted, and that natural resources are to be left to nature—just as Marx did.

Marx was horribly wrong, and so are they.

The U.S. is currently at a resource crossroads. Census figures reveal that in forty years there will be an additional 100 million residents in our country. We need a rational energy policy, more water, additional resources such as timber and minerals, and more land for farming and living. However, if a lifestyle that includes rationed electricity, severe water shortages, high-density housing, higher gasoline prices, and crowded mass-transit systems sounds good to you, then don't bother reading the rest of this book.

In the following chapters, we will discover how expansive America's natural resources are, how the environmentalists—led by the Obama administration—are attempting to lock up those resources, and how we can responsibly harvest our nation's uncultivated wealth to further America's greatness into future generations.

PART TWO

SEVEN

SOLAR AND WIND: THE INCONVENIENT TRUTH

ON MAY 26, 2010, as one of the largest oil leaks in history poured into the Gulf of Mexico, President Barack Obama read from his prepared notes and drilled Big Oil: "Part of what's happening in the Gulf is companies are drilling a mile underwater before they hit ground, and then a mile below that before they hit oil. With the increased risks, increased costs—it gives you a sense where we're goin'. We're not going to be able to sustain this kind of fossil fuel use. This planet can't sustain it."[1]

One would have supposed his remarks were made from the shores of Louisiana, but no, they were issued from the site of one of the biggest taxpayer-funded financial disasters in the history of energy—alternative energy, that is—the headquarters of Solyndra, a solar panel manufacturing firm located in the Silicon Valley.

The company had received a guaranteed loan of more than half a billion dollars, with terms that included a quarterly interest rate of a ridiculously low 1.025 percent.[2] Obama was visiting the plant to witness how the money had been spent and to solicit adoration from Solyndra investors, employees, and media hacks who weren't interested in digging into the real story that was right under their noses.

"It's here that companies like Solyndra are leading the way toward a brighter and more prosperous future," Obama proclaimed.

While Obama went on to highlight the company's "cutting-edge solar panels," insiders knew the firm's proprietary technology was actually "bleeding edge." In other words, the product was so unique there was a high risk of it being unreliable and more expensive to employ than it was worth. Potential investors had been avoiding Solyndra for some time, and current shareholders were not reopening their checkbooks.

Truth is, during the last days of the Bush administration, the Department of Energy and the independent Office of Management and Budget unanimously decided to take a pass on making a loan commitment to Solyndra. Outside credit agencies, such as Dun & Bradstreet, assessed it as "fair." Projections were that the company had no chance of getting out of the red for many years, if ever. In the real world, such evaluations are not followed with shelling out $535 million at low interest. But proving that Team Obama is not about reality, the loan was finalized using funds found within the $800 billion American Reinvestment and Recovery Act, a.k.a. the Stimulus, in April, 2009.

Announcement of the deal was made at Solyndra's headquarters. It was a mega-media event that included Governor Arnold Schwarzenegger, Energy Secretary Steven Chu, and Vice President Joe Biden, who appeared via satellite.

"By investing in the infrastructure and technology of the future," Biden proclaimed, "we are not only creating jobs today, but laying the foundation for long-term growth in the twenty-first-century economy."[3]

Secretary Chu, whose department sponsored the loan, stated, "This investment is part of a broad, aggressive effort to spark a new industrial revolution that will put Americans to work."[4]

Obama, Biden, and Chu, however, would all be proven horribly wrong.

You must realize that by the time the Obama White House got involved, Solyndra was the Silicon Valley's most well-capitalized start-up company *ever*, having raised more initial funding than high-tech titans Google, Amazon, and Apple. Conversely, Solyndra was spending more money than any start-up. Since 2005, the firm had burned through nearly a billion dollars of private equity, using a substantial portion of it to build an exquisite 183,000-square-foot facility staffed with nearly 1,000 employees. Once the Stimulus cash was tapped, the company rushed to develop an additional 609,000 square feet of manufacturing space, including shipping docks, office suites, an expansive staff cafeteria, and a 4,000-square-foot employee gym.[5] More workers were also to be hired with the easy money. "When it's completed in a few months," Obama declared during his 2010 visit, "Solyndra expects to hire [an additional] 1,000 workers to manufacture solar panels and sell them across America and around the world."[6]

A month after the president's high-profile visit to Solyndra, the panel maker had planned for an Initial Public Offering of stock, but with little notice those plans were surprisingly scrapped. In July, its CEO and founder, Chris Gronet, suddenly quit. Obama's green darling was in serious trouble, and it didn't require a degree in economics to figure out what was next.

Curiously, on the day *after* the 2010 midterm elections, Solyndra made public that it would shut down its original manufacturing facility, lay off some 200 workers, and cancel plans for the 1,000 new jobs that Obama had talked up just a few months prior.

You have to wonder who made the call for the date of this announcement? All of the polling had indicated that the Democratic Party was in for a shellacking on Election Day. A preelection front-page story detailing a half billion dollars in aid to a green company endorsed by the president wouldn't help Democratic candidates in tight races. In any case, when the story finally broke, it was buried beneath the buzz over the Tea Party taking the House of Representatives and hardly made news in California. From a PR perspective, the announcement's timing was extremely clever.

However, I soon composed a national story for *Human Events* on the Solyndra debacle, and Republican members of Congress took note. In February 2011, the Department of Energy loan to Solyndra came under scrutiny, as the chairman of the House Energy Committee, Congressman Fred Upton of Michigan, requested documents from the Department of Energy about the loan guarantee, raising questions as to whether it was a prudent investment choice. Soon, other media outlets jumped in, most notably the *Wall Street Journal*, who asked Solyndra's new CEO, Brian Harrison, about the ARRA money. "The $535 million was not a grant, not a subsidy, it was a loan," Harrison said, adding that the company fully intended to repay it.[7]

Asked whether the company would have taken the federal loan, given all the scrutiny, in hindsight Harrison said, "I suppose we would do it again." He also stated that the loan enabled the company to build their new factory, which turned out to be the second largest construction project in California in 2010.

Mr. Harrison was uttering eco-speak. Once again, U.S. taxpayer money was being spent for U.N.-inspired "green" construction jobs.

If you were to see Solyndra's facility in the city of Fremont, just north of San Jose and south of Oakland, you would agree it's the most over-the-top manufacturing plant design ever. Many suspect the site was chosen as a political payback to longtime allegiant Democrat, Congressman Pete Stark, whose district includes Fremont. Stark, by the way, is a caustic, crusty old guy known for cussing out constituents at town hall events. He is also the toast of the anti-war movement who even once suggested on the House floor that President Bush was sending troops to Iraq to get their "heads blown off for his amusement."[8] Stark has always been an outspoken union advocate and all the labor required for the Solyndra build-out was completely supplied by high-cost unionized firms. If my speculation is correct, it may explain the wild rush to finish the second, larger phase of the campus. Crews were working 24/7 and racking up tremendous overtime wages, with construction costs for the excessive facilities estimated at well over twelve hundred dollars per square foot.[9]

Beyond the building expenditures, because Solyndra's solar panels were being manufactured in an area known for its extreme cost of living and incredibly restrictive environmental policies, the price of getting the product to market soon proved to be completely unsustainable.

Solyndra was upside down—despite what appears to be significant efforts by investors to take advantage of their cozy relationships with the White House.

GREEN CRONYISM

One of those investors was George Kaiser, a man who personally gave Obama's presidential campaign $53,500 and did some heavy-duty fund-raising for the Democrats as well.[10] The George Kaiser Family Foundation, a charitable organization based in Tulsa, Oklahoma, held about 35.7 percent of the company, according to a Solyndra filing with the Securities and Exchange Commission. While the foundation insists "George Kaiser is not an investor in Solyndra and did not participate in any discussions with the U.S. government regarding the loan,"[11] it is interesting to note that Kaiser was responsible for sixteen of the twenty meetings that showed up on the White House logs associated with Solyndra associates.[12] A month before the loan deal was announced, Kaiser had meetings on March 12 with former chairman of the Council of Economic Advisors Austan Goolsbee at 11 a.m., senior advisor Pete Rouse at 3 p.m., and deputy director of the Domestic Policy Council Heather Higginbottom at 6:30 p.m. The following morning, Kaiser met with deputy director of the National Economic Council Jason Furman at 9.

Cracks in the Kaiser Foundation's story appeared when an email sent by George Kaiser on March 5, was released by the House Energy and Commerce Committee. In the communiqué, Kaiser said that when he and a foundation official visited the White House the previous year officials showed "thorough knowledge of the Solyndra story, suggesting it was one their prime poster children" for alternative energy.

In another March 5 email, a well-placed Kaiser associate appears sure that Energy Secretary Steven Chu would approve a second loan for Solyndra.

"It appears things are headed in the right direction and Chu is apparently staying involved in Solyndra's application and continues to talk up the company as a success story," wrote Steve Mitchell, managing director of Kaiser's venture-capital firm, Argonaut Private Equity. Mitchell also served on Solyndra's board of directors.

White House logs reveal other Solyndra officials that popped into the White House, including founder Christian Gronet on September 22, 2009, at 9:30 a.m.; board member Thomas Baruch visited May 7, 2010, and September 20, 2010, at 8:40 a.m. and 1 p.m., respectively; and David Prend, another board member, visited on September 21, 2010, at 9:15 p.m.

I doubt they stopped by to shoot hoops with the prez.

Another Solyndra investor close to Obama is former California controller Steve Westly, CEO of the capital firm the Westly Group. He was a big-time money-bundler for candidate Obama in 2007–08, raising a half million dollars for the campaign. In August 2010, Westly was rewarded with a seat on the White House's Energy Advisory Board. Three months later, Westly received the president as a guest in his home as part of an exclusive fundraiser for Obama's friend, state attorney general candidate Kamala Harris. Westly was also one of the chosen few who supped with Obama at an intimate February 2011 dinner in the Silicon Valley home of Al Gore's investment buddy John Doerr. Other select fellows included Steve Jobs of Apple, Eric Schmidt of Google, Dick Costolo of Twitter, and Mark Zuckerberg, founder of Facebook.

Westly's special relationship with Obama has certainly paid off, as four companies in his investment portfolio have secured hundreds of millions in federal loans and grants.[13] As of September 2011, his company's website boasted, "We believe that with the Obama administration, and other governments in Western Europe and Asia, committing hundreds of billions of dollars to clean tech, there has never been a better time to launch clean tech companies. The Westly Group is uniquely positioned to take advantage of this surge of interest and growth."[14]

Despite the distinct tones of green cronyism, it all came crashing down for Solyndra on August 31, 2011. Shortly after 5 a.m., an anonymous caller to my radio program told me on-air that all employees of the firm would be fired that morning at 6. The insider was spot-on. The company released everyone, closed its doors, and filed Chapter 11. We the people were burned to the tune of a half billion dollars. Further illustrating their utter tone deafness, a day after Solyndra's failure, the Department of Energy awarded another $145 million to sixty-nine solar projects taking place at universities, government research labs, and major corporations.[15]

With the solar green giant down for the count, email communiqués were released by the House Energy and Commerce Committee revealing that bean counters within the Obama administration were hesitant to go through with the loan. "This deal is NOT ready for prime time," warned one White House budget analyst in a March 9, 2010, email, written nine days before the loan was formally announced. On March 7, Ronald Klain, who was chief of staff to Vice President Biden, wrote, "If you guys think this is a bad idea, I need to unwind the W[est] W[ing] QUICKLY."

Emails from August 2009, a month before the big Biden satellite event, reveal an official at the OMB complaining about "the time pressure we are under to sign-off on Solyndra." Another cited, "There isn't enough time to negotiate."

Actually, that August was a revealing month for solar energy. In addition to Solyndra going under, a Massachusetts corporation, Evergreen Solar, announced it would file for bankruptcy after receiving nearly $60 million in state funding, and benefited from several major installations that were paid for with Stimulus dollars.[16] Evergreen's website actually boasted about how the Stimulus money was going to keep their company vital, "As the number of ARRA solar projects continue to increase across the U.S., our ES-A series panels will continue to be the panel of choice for American-based installers and end users."[17]

Another manufacturer, SpectraWatt of New York, filed for bankruptcy after cashing in on public support totaling nearly $8 million, $500,000 of which came from a National Renewable Energy Laboratory (NLRE) grant. The NLRE money was made of Stimulus funds.

Also in August, one of the world's largest solar manufacturers, First Solar of Tempe, Arizona, greatly reduced its profit estimates after being forced to lower prices and incur higher expenses—not a promising recipe for success.

On Solyndra's final day, Reuters news service sobered up to solar, noting, "Even industry heavyweights such as China's Suntech Power Holdings Co. Ltd. and U.S.-based First Solar, Inc. are struggling with dwindling profits, while small, up-and-coming solar companies are finding it increasingly difficult to stay afloat."[18]

While harnessing the energy of the sun, and for that matter, the wind, seems like such an altruistic endeavor; the inconvenient truth is these sexy-sounding alternatives are expensive, inefficient, require lots of real estate, and necessitate that a backup source of power be available at all times—generally in the form of fossil fuel.

PRICEY BY THE WATT

As opposed to electricity produced by natural gas, coal, nuclear, or hydro, solar power is incredibly expensive to produce and, therefore, pricey to the consumer by the watt.

A watt is a measure of electricity used by a power-consuming device. A thousand watts are known as a "kilowatt." A kilowatt-hour (kWh) is the amount of energy consumed when a kilowatt is expended or consumed over the course of an hour. Your utility company uses kilowatt-hours as the basis for what they charge you.

For example, a space heater rated at 1,000 watts operating for one hour uses one kilowatt-hour of electricity. If a 46" flat-screen TV uses 100 watts when it's turned on, then over the course of ten hours, it will consume one kilowatt-hour of power. An average electric clothes dryer is rated at about 4,000 kWh. That means it's consuming a kilowatt hour every fifteen minutes. So, if your utility company is charging you the national average retail price of 11 cents per kWh, running that dryer for sixty minutes will cost you 44 cents.[19] Easy enough.

Now, here's where it gets a bit more complex.

To arrive at the price of a kilowatt-hour of energy, your utility company looks at the various energy sources in their portfolio and carefully determines the cost involved to bring each product to market. This is called the "production cost." Their calculations include everything from the price of raw materials (oil, natural gas, or nuclear fuel rods), the construction costs of the power plant, ongoing maintenance, personnel, administration, etc. After hydroelectricity (which we will address in chapter 11), nuclear power and coal are the cheapest sources, with production costs averaging 2.14 cents for nuclear and 3.06 cents per kWh for coal; natural gas is next at 4.86 cents. Three percent of the electricity in the U.S. is generated from petroleum, which costs 15.18 cents to produce.[20]

Your personal per-kilowatt hour cost will depend on the blend of energy sources running through the power lines to your house, as well as taxes, fees, and surcharges.

Obviously, if your home receives all its electricity from nuclear or coal, your kWh charge is likely dirt-cheap, and in most locations across the country, you're able to run a dryer on a hot afternoon in July without so much as a second thought. However, if, perchance, your utility supplier is sending you electricity generated 100 percent by solar, you would stop using the automatic clothes dryer forever and hang your Skivvies on a line in your backyard, to be sun-dried. Why?

Solar energy is an amazing technology—about as close to alchemy as one can get—but it's virtually impossible to accurately price. We do know this: it's very expensive.

Let's examine the most popular form of solar energy, photovoltaic.

Photovoltaic solar was discovered in 1839 by French physicist Edmund Bequerel, who realized that certain materials would produce small amounts of electric current when exposed to light; this became known as the "photovoltaic effect." In 1905, Albert Einstein accurately detailed the nature of light and the photoelectric effect on which photovoltaic technology is based, for which he later won a Nobel Prize in physics.

The first photovoltaic module was built by Bell Laboratories in 1954 and was mostly considered a novelty, as it was far too expensive to gain widespread use. In the 1960s, the U.S. space program began to make the first serious use of the technology to provide power aboard spacecraft. Through the space program, photovoltaic solar power advanced, and its reliability was established, though costs remained astronomical. In 1979, an oil shortage occurred in the wake of the Iranian Islamic Revolution, and American eco-socialists used the crisis to demonize the fossil fuel industry and encourage the use of solar power for non-space applications.

It was yet another fanciful progressive illusion.

Photovoltaic solar is a big-ticket proposition. Besides consisting of materials you can't purchase at the hardware store—monocrystalline silicon, polycrystalline silicon, amorphous silicon, cadmium telluride, and copper indium gallium selenide—there is the complex design of the high-tech panels, plus the steep assembly costs. Once manufactured, there is the installation process. Since the sun travels across the sky at different trajectories throughout the year, the panels need to be positioned at the best possible angle to catch the most direct radiation. If you want to utilize panels that tilt with the sun's passing in order to better capture direct light, that will increase efficiency and greatly add to the bottom line. Next (it sounds trivial, but it's important) the panels need to be kept clean—a thin layer of dust or droppings from passing birds will decrease the panels' energy-producing capacity. And then there is degradation. The sun is a terribly harsh master, and as solar panels weather, their efficiency is further reduced.

It's so much easier to dig up a hunk of coal and toss it into a lit oven—or ignite some natural gas—and use the cheap fuels to boil some water, enabling the associated steam to drive an electricity-generating turbine.

For solar panel manufacturers, the obvious holy grail is getting the kWh cost down to something near traditional energy sources. Right now it's not even remotely close—in fact, it's nearly impossible to quantify the real cost of solar. The problems include the fact that traditional sources of energy are available 24/7—solar energy obviously is not. Dark of night, clouds, and angle of the sun deter efficiency, thus making the power source more expensive to produce. Plus there's an additional quagmire for solar that most promoters do not like to talk about: its noted limitations require a full-time backup system, continually running at a low level, in order to be able to quickly ramp up in the event of a change in atmospheric conditions. Most such backups are natural gas–driven.

Is there currently a practical use for solar power? Sure. Photovoltaic solar is often a wise choice for powering small-scale applications such as street lighting, road signage, landscape illumination, and water pumping; it's wiser still for bringing electricity to remote, off-the-grid locations, where other options are impossible or too costly. In the meantime, for large-scale usage, it remains a technology that needs to be further developed by the private marketplace, not pushed by bureaucrats and slick politicians. Instead, we have a government-sponsored mad rush to get the product to market when it's " NOT ready for prime time." And, given how high tech is always improving (your home computer becomes a relic in just seven years), the panels currently being installed on gargantuan "solar farms" today will be sorely outdated in a just a few years.

And then there is the land required for these farms.

SOLAR'S REAL-ESTATE REQUIREMENT

In 2010, the largest solar project of its kind went online in California's Central Valley. The solar array, known as CalRenew-1, is located in the small town of Mendota, and is owned by Meridian, a New Zealand company. The plant uses made-in-Japan solar panels, and covers 50 acres of some of the richest farmland on the planet. The real estate is being leased to Meridian in a sweetheart deal brokered with the City of Mendota for $833 per month. The plant produces a measly five megawatts of power that Meridian is selling to the region's largest energy provider, Pacific Gas and Electric.

Five megawatts of solar power is a puny amount of energy in the utility world, roughly enough to briefly power 2,500 homes (during the peak of the afternoon, in the summer, assuming it's perfectly sunny). The real goal of the plant, though, was to meet a mandate set by the State of California, demanding that renewable energy (wind, solar, geothermal, biomass, and small hydro) make up 20 percent of the electricity sold in the state.

In early 2011, a new law was passed by the ultra-green-leaning legislature demanding that the amount be raised to 33 percent. The new law was passed despite a report in 2009 from the California Public Utilities Commission saying that energy costs would likely increase 7.1 percent if a third of electricity came from renewable sources.[21] In a dash to adhere to an agenda, the lawmakers decided to disregard the consumer and move ahead with their green plan.

And because environmentalists in California are so opposed to expanding the use of fossil fuels, CalRenew-1 avoids the need for a natural gas backup system because its only job is to provide energy to the electrical grid during times of peak usage, primarily in the summer months.

Now, compare the real estate footprint of the Mendota plant (50 acres) to that of a small, natural gas facility generating forty times as much electricity (200 megawatts) and serving 100,000 homes. One such plant is set for construction in Eklunta, Alaska. It will require about 70 acres of land. Doing the math, that means to generate the same amount of energy as the proposed Eklunta plant, CalRenew-1 would require 2,000 acres.

Next, compare solar's real estate requirement to that of a significant nuclear power plant, the Diablo Canyon facility. This California plant produces 2,200 megawatts of emission-free energy and sits on 800 acres. Comparably, a field full of solar panels equaling the amount of power coming out of Diablo Canyon would amount to 22,000 acres, or about 34 square miles—nearly the size of San Francisco. That's unrealistic.

Not to step on toes here, but solar power is reminiscent of a school science project funded from a benevolent parent's checking account. Thousands of tinkerers working with federal grants, government loans, university research money, and venture capital are trying to compete with what we know already works very well: fossil fuels, hydropower, and nukes.

At first glance, solar energy looks extremely promising, but without government subsidies, it doesn't warrant a second look.

SUBSIDIZING SOLAR

An increasing number of homeowners across the nation have decided to make major investments in rooftop solar systems. For some the choice is a matter of wanting to see their electricity bills shrink; for others it is a public declaration of being green. No matter what the motivation for installing rooftop solar, though, the decision is all the easier because of the fat subsidies provided by federal and state taxpayers. More than half the states in the U.S. offer enough incentives to cut the costs by 40 percent or more, according to the Database of State Incentives for Renewables and Efficiency.[22]

For example, prior to Chris Christie becoming governor of New Jersey, a variety of whopping incentives were available to pay for more than 90 percent of home solar systems. A 5-kilowatt rooftop array dropped from $37,500 to about $2,625 after applying the federal tax credit, a state rebate, and a renewable energy program through the state's largest electric utility, PSE&G. The initial rationale for the immense giveaways was that home electricity consumption would drop 40 percent, saving homeowners hundreds of dollars annually, and creating lots of green jobs in New Jersey. With the artificially low costs being offered to the homeowner, the outlay required to purchase the array could be recouped in just a few years. The whole program was a nice-sounding idea, but was financially unsustainable and, in the case of the Garden State, was just one of many do-gooder schemes that caused it to sink into deep fiscal doo-doo.

One of the first things Governor Christie did after taking office in 2010 was end the state rebate solar program.

Solar's surge, both domestically and abroad, is the sole result of bureaucratic subsidies. In 2010, the Treasury Department had $2.3 billion in tax credits available for companies that made energy equipment, including solar panels. Solar companies have also competed for $11.3 billion in money derived from the Stimulus act, and in 2008, Congress agreed on $2.5 billion in tax credits for homeowners installing solar over the next ten years.

Other notable solar handouts from Secretary Chu's department include a $1.2 billion loan guarantee for the California Valley Ranch solar farm owned by NRG Energy and built by SunPower, and $1.4 billion to Abengoa Solar for a solar plant in Arizona. The Obama-supporting guys at

Google also scored some of the green Stimulus cash. Google invested $168 million in BrightSource, a solar farm located in the Southern California desert. BrightSource received a whopping $1.6 billion from the DOE.[23]

Those dinners with Obama can really pay off.

For example, next door to Solyndra (and still in Congressman's Stark's district) is the gigantic Tesla Motors plant. Tesla is making all-electric cars (there are currently two models under production—one priced at over $100,000, the other at just over $54,000). Tesla took a $465 million federal loan.[24] One of Tesla's key investors is Mr. Westly.

Then there's Fisker Automotive in Finland; they're making another pricey electric car. Fisker is part-owned by John Doerr and Al Gore's investment firm, Kleiner Perkins. And guess who scored $529 million of Stimulus cash? Fisker Automotive.[25]

It pays to rub shoulders with the proper people.

Even our own U.S. Energy Information Administration—the independent number-crunching wing within the Department of Energy—agrees that the only thing keeping solar power afloat is the subsidies. Their data illustrates that total federal subsidies (including everything from tax breaks to research grants) for solar power were $24.34 per megawatt hour, compared to 44 cents for traditional coal, 25 cents for natural gas and petroleum liquids, 67 cents for hydroelectric power, and $1.59 for nuclear.[26] Wind power, which we will address in a moment, is subsidized at almost as high a level as solar, at $23.37 per megawatt hour.

Examining the price kickbacks for rooftop solar state by state reveals a lot about how states are governed. According to the Associated Press, here are the sweeteners offered to buyers of an average (5-kilowatt) residential system in several states:[27]

- New York. The hefty $40,250 retail cost is slashed after applying a federal tax credit of $8,243, a state tax credit of $5,000, a city property tax abatement worth $2,404, and other local incentives worth $12,775. Final price: $11,828.

- Massachusetts. The initial $36,500 price tag gets cut after applying a federal tax credit of $9,150 and state incentives totaling $6,000. Final price: $21,350.

- California. The original $40,000 sticker price would be cut after applying a federal tax credit of $9,690 and a rebate of $7,700 through Southern California Edison. Final price: $22,610.

- Arizona. The $35,000 cost before incentives is reduced by the federal tax credit of $6,000, a state rebate of $15,000, and an Arizona tax credit of $1,000. Final price: $13,000.[28]

- Arkansas. The big $50,000 price tag is cut by $15,000 after applying a federal tax credit. Final price: $35,000. Not the best state to scam the system.

After reviewing the kickbacks and financial incentives to go solar, it's no wonder our governments are financially in ruins.

Now witness the 2011 budget deficits of these same states offering solar handouts:

- New Jersey, $11 billion
- New York, $10 billion
- Massachusetts, $3 billion
- California, $26 billion
- Arizona $2 billion
- Arkansas and Alaska, *no deficit*

Solar is a wonderful technology, but it's a pretentious power source that, for major applications, is only affordable when the government is shelling out freebies or, as is the case in California, when its use is mandated by law. I personally have used solar to heat my swimming pool and supplement my water heater, but that's a passive form of solar that is very inexpensive to install, and extremely cost-effective.

Expansive loan guarantee programs are fraught with problems. At a minimum, they create taxpayer liabilities, give recipients preferential treatment, and distort capital markets. Further, depending on how they are structured, such programs stifle innovation, remove incentives to decrease costs, suppress private-sector financing solutions, perpetuate regulatory inefficiency, and encourage continued government dependence.

Limited government loan guarantees can help overcome some near-term financing obstacles, but they are subsidies. If not used with the utmost prudence, such financing programs only act to prop up noncompetitive industries, magnify market uncertainty, and make private financing difficult to attain—and after all, private financing has the ability to create wealth, jobs, and therefore, tax revenue.

However, as one would expect, the Obama administration has a track record of turning a blind eye to free-market principles. In June 2011, Secretary Chu announced two expensive new solar "contests" that will cost the taxpayers $28 million.

Chu's "Solar Rooftop Challenge" will hand out $13 million to as many as twenty-five local government agencies that are able to best streamline the permitting process for rooftop solar installations. His "Balance of System Costs" program will dish out $15 million to local governments who can do the same for utility providers. The official DOE press release states that the funding will provide "a critical element in bringing down the overall costs of installed solar energy systems."[29]

I disagree. Government largesse simply perpetuates the systemic inefficiencies and risk that gave rise to the need for subsidies in the first place.

WIND'S NO BETTER

It's absurd that our own government officials won't lend an ear to the folks at the taxpayer-funded Energy Information Administration. In the words of the EIA, solar and wind are not worth the money we are throwing at them:

> Wind and solar are intermittent technologies that can be used only when resources are available. Once built, the cost of operating wind or solar technologies when the resource is available is generally much less than the cost of operating conventional renewable generation. However, high construction costs can make the total cost to build and operate renewable generators higher than those for conventional power plants. The intermittence of wind and solar can further hinder the economic competitiveness of those resources, as they are not operator-controlled and are not necessarily available when they would be of greatest value to the system.[30]

Not a strong sell to go green. And regarding wind, what the EIA didn't tell you was a very real complaint that's making wind turbines an obnoxious alternative—their sound.

"Wind One" is the turbine owned by the town of Falmouth, Massachusetts. The thirty-two thousand residents of Falmouth almost universally welcomed Wind One as a symbol of renewable energy and a way to keep taxes down and save the town about $400,000 a year in electricity costs. Located near the town's wastewater treatment plant on the southwestern tip of Cape Cod, the turbine's blades extend just shy of 400 feet. Given that the topography of the area is beautifully enhanced with small ponds, creeks and inlets, and pines and oaks adjacent to the rocky beachfront, witnessing the monstrous pillar of white steel and its whirling three-bladed rotor is at first shocking. As one resident told public radio station WGBH, "It's quite majestic."[31]

Installed in the spring of 2010 at a cost of $4.3 million (with $3 million paid through the Clean Energy Program/Community Wind Collaborative, grants and renewable energy credit purchase), the huge turbine cranks out 1.65 kilowatts of electricity in optimum conditions. The windmill would be dedicated to powering the wastewater facility.

But there was a problem. Have you ever heard the chopping sound associated with the blades of a large helicopter? That's the noise Wind One blasts anytime a breeze reaches about twenty miles per hour—which is very common on the Cape.

As soon as Wind One went online, residents began to complain.

Neil Anderson lives a quarter of a mile from the windmill. He's an avid supporter of alternative energy, having owned and operated a passive solar company on Cape Cod for the past twenty-five years. "It is dangerous," he explained. "Headaches. Loss of sleep. And the ringing in my ears never goes away. I could look at it all day, and it does not bother me—but it's way too close."[32]

More and more residents complained about Wind One, and eventually a group lawyered up. A deal was struck with the town of Falmouth to disengage the turbine when winds exceed twenty-three-plus miles per hour. But this is problematic because large windmills such as Wind One operate at optimum efficiency at about thirty miles per hour.

So, Falmouth's investment has taken a big hit. According to Gerald

Potamis, who runs the wastewater site, shutting off the turbine during periods of high winds will cost the town $173,000 in annual revenue, because now they'll have to use more traditional power (like they do when the wind is calm, or when the wind speed is too high to operate safely).[33]

And it will just be a matter of time before the turbine takes another hit—from lightning. The massive steel structures are lightning magnets. Go to YouTube, search "lightning strike wind turbine," and you'll see quite a few dramatic displays of the turbines getting fried.

All said, wind turbines suffer from the NIMBY syndrome—*not in my backyard.* Look no further than the largest concentration of wind turbines in the world, constructed in the 1970s just east of the San Francisco Bay. Some 4,500 windmills are ensconced atop 50,000 acres of grassy hills, generating a modest 576 megawatts of power. Officially known as the Altamont Pass Wind Resource Area, the wind farm could be presumed to be an icon of greenness. But instead, Altamont Pass is the poster girl of eco-infighting.

I've been driving by this wind farm for years. No one really lives nearby, so the sound, which replicates a massive helicopter invasion, is not a bother to folks—but the animal rights people hate this array and have been trying to shut it down ever since it went online.

Ever since the multitude of three-bladed rotors was installed, a significant increase in the numbers of dead birds in the area was reported. Activists immediately went ballistic, demanding action. Since then, lawsuits have been filed and millions of dollars spent procuring studies to track the bird body count to determine how to address the problem. In 2008, the most extensive study was released: a two-year, taxpayer-funded examination conducted by the Altamont Pass Avian Monitoring Team. It surveyed 2,500 of the turbines and kept meticulous records of the bird body count. During the study period, 1,596 rotary-blade bird deaths were confirmed, including the deaths of 633 raptors.[34] Extrapolating their data to account for all 4,500 windmills on the farm, as well as estimates regarding how many dead birds the researchers didn't count before scavengers made off with an easy meal, the monitoring team claims that in two years 8,247 birds died of what I refer to on the radio as "turbine to the head syndrome."

In 2010, a settlement had been reached between the Audubon Society, Californians for Renewable Energy, and NextEra Energy Resources (who operate some 5,000 turbines in the area). Nearly half of the smaller turbines will be replaced by newer, more bird-friendly models. The project is expected to be complete by 2015 and includes $2.5 million for raptor habitat restoration, all of which is expected to increase the price of energy being supplied by the wind farm.

BREAKING WIND MYTHS

Wind currently provides about 1.8 percent of the electricity generated in the United States. This amount has been increasing by a fraction of a percent each year for the last several years. Texas generates the most wind power of any state, more than 14 billion kilowatt-hours; Iowa generates the most power by wind as a percentage of all the electricity generated in the state, 14.2 percent.

Wind power is loud, unseemly, inefficient, and costly to the taxpayer. A counter argument often presented by environmentalists is "the subsidies to Big Oil dwarf that of wind and solar."

In the next chapter, I will illustrate Big Oil's profits and the amount of money it pays in taxes, but for now, let's break down wind power and total up the tax breaks and other forms of subsidies in the alternative energy sector. Again, we turn to data from the EIA. As mentioned previously, solar is subsidized at $24.34 per megawatt hour, and wind at $23.37.

For the true believers who claim wind and solar need more assistance, may I remind you that wind and solar have been riding the subsidy wagon for years, and yet still account for less than 3 percent of total net electricity generation.

By comparison, nuclear power provides 20 percent of U.S. base electricity production, yet it is subsidized about fifteen times *less* than wind. Traditional coal supplies power to nearly half the nation's grid, and its subsidies are fewer still. And if we want to talk about the number of people permanently employed in these industries, there is no comparison; although recently the *Huffington Post* ran a blurb that compared employment in wind versus the coal industry—a blurb that was way off the mark of truth: "The wind industry now employs more people than coal mining in the United States. Wind industry jobs jumped to 85,000 in 2008, a 70 percent increase from the previous year, according to a report released

Tuesday from the American Wind Energy Association. In contrast, the coal industry employs about 81,000 workers."[35]

The Wind Energy Association is numbering anyone who even *thinks* about a wind turbine, from the designers to the manufacturers to the installers and maintenance guys. For the coal industry, the figures only include the miners. A genuine method for counting those employed in the coal industry would be to include everyone involved, from the dark tunnels of the mines to power plant. That figure will produce a full-time employment number over 1.5 million.[36] If you are looking for an energy source that provides both plentiful jobs and inexpensive electricity, it's hard to beat coal—but of course, the environmentalists demonize coal for its association with carbon dioxide emissions, and the materialists are against making a profit off a natural resource. To quote Al Gore, "If you're a young person looking at the future of this planet and looking at what is being done right now and not done, I believe we have reached the stage where it is time for civil disobedience to prevent the construction of new coal plants."[37]

However, it's interesting to note that wind power doesn't necessarily reduce the user's carbon footprint.

In comments made to Ontario, Canada's, legislature regarding a bill involving renewable energy, Michael J. Trebilcock, a professor of law and economics at the University of Toronto, said,

> There is no evidence that industrial wind power is likely to have a significant impact on carbon emissions. The European experience is instructive. Denmark, the world's most wind-intensive nation, with more than 6,000 turbines generating 19 percent of its electricity, has yet to close a single fossil-fuel plant. It requires 50 percent more coal-generated electricity to cover wind power's unpredictability, and pollution and carbon dioxide emissions have risen (by 36 percent in 2006 alone).
>
> Flemming Nissen, the head of development at West Danish generating company ELSAM (one of Denmark's largest energy utilities) tells us that "wind turbines do not reduce carbon dioxide emissions." The German experience is no different. *Der Spiegel* reports that "Germany's CO_2 emissions haven't been reduced by even a single gram," and additional coal- and gas-fired plants have been constructed to ensure reliable delivery.[38]

What a dilemma for the greenies. The problem is wind—and solar—may actually *increase* carbon dioxide emissions in some cases, depending on the carbon-intensity of backup generation required because of its intermittent performance. Such sources of power are called "baseload," as opposed to peak demand generating facilities that can ramp up and down when needed.

Once battery storage exists for wind and solar (which will be incredibly expensive), wind and solar could be considered baseload—but don't hold your breath.

A nuclear engineer shared with me an example involving the "Prosperity Energy Storage" demonstration project in New Mexico, funded in part with $2 million Stimulus dollars. The program will utilize 2,158 solar panels to generate 500 kilowatts of electricity, with 1,280 batteries storing 250 kilowatts of that power—enough to power 150 average homes for two hours each day. Based on those figures, my engineer friend said we could have battery storage to power 12.5 homes for twenty-four hours... at a cost of $696,000 per home!

In the meantime, wind and solar will continue to be pushed by politicians wanting to create green jobs, do-gooders desiring to save polar bears, and committed eco-socialists who do not believe anyone should be able to profit off of raw natural resources such as coal, natural gas, and oil.

EIGHT

WE OWN IT—LET'S DRILL IT

ON MARCH 11, 2011, the day Japan was slammed with a monster 8.9 earthquake and subsequent tsunami, President Barack Obama held a news conference to address the tragedy. After a few brief remarks regarding the catastrophe, he suddenly turned and unleashed on those critical of his domestic *oil* policy.

The timing for such a tongue-lashing was beyond awkward.

"Now, before I take a few questions, let me say a few words about something that's obviously been on the minds of many Americans here at home, and that's the price of gasoline."

Talk about detachment. Tens of thousands of people had just been killed, a nuclear reactor was severely damaged, and illustrating the tsunami's awesome power, its leading wave had reached the West Coast of the United States, causing $50 million worth of damage to two harbors in California.

Yes, gas prices were rising at a record clip, but a scripted lecture aimed at the drill-baby-drill club, now?

"Any notion that my administration has shut down oil production might make for a good political sound bite, but it doesn't match up with reality. We are encouraging offshore exploration and production. We're just doing it responsibly. We've got to work together—Democrats, Republicans, and everybody in between—to finally secure America's energy future. I don't want to leave this for the next president, and none of us should want to leave it for our kids."

Not only did Obama choose the wrong moment to defend his energy policy, but his brash assertions did not align with reality. The president was guilty of subterfuge. Let's set the record straight.

On May 6, 2010, three weeks after the onset of the massive British Petroleum (BP) oil spill in the Gulf of Mexico, the Obama administration cleverly took advantage of the crisis by autocratically imposing a thirty-day moratorium on new deepwater drilling operations in the Gulf, and suspending drilling on rigs working in water deeper than five hundred feet. This mandate immediately froze operations on thirty-three operational oil platforms, and another eight that were under construction, pending an investigation into the explosion that generated the BP leak.[1]

But the Obama administration didn't just limit its antidrilling dictates there. Simultaneous to the Gulf abeyance, the administration also suspended applications for exploratory drilling in the Alaskan Arctic until an unspecified date in 2011. This extreme action was taken despite the fact that in 2009, preliminary permits had been previously issued by the Interior Department to Shell Oil to drill five wells in the region—three in the Chukchi Sea and two in the Beaufort Sea.[2] The suspension was a shock to Shell, who had been planning the projects for the better part of a year. They were planning to begin drilling operations in June, as the summer weather would allow them a brief four-month window of opportunity to work, but the Obama administration's decision shot that plan down for at least another year.

However, Team Obama didn't stop there. Prior to the expiration of the thirty-day Gulf drilling moratorium, on May 27 the stay was extended for an extra six months. In addition, the president also ordered that the Atlantic Coast be off limits to energy development or exploration through 2017. This was a stunning reversal of the promise Obama made to the people of Virginia just two months earlier, on March 31, when he announced the lease-sale of nearly 3 million acres off their coast, a move that would enable America to harness some 130 million barrels of oil, and more than a trillion cubic feet of natural gas. According to Virginia governor Bob McDonnell, the offshore projects would "speed our economic recovery."[3] McDonnell estimated that the plan would have added 2,578 full-time jobs annually, induce capital investment of $7.84 billion, yield $644 million in direct and indirect payroll, and result in $271 million in new state and local revenue.

It was an opportunity lost for creating real American jobs.

Meanwhile, the Gulf freeze was challenged in federal court twice, with the administration losing both times. The first loss occurred on June 22, 2010, when U.S. district judge Martin Feldman in New Orleans stated that the Interior Department "acted arbitrarily and capriciously" when it incorrectly assumed that because one rig failed, all companies and rigs doing deepwater drilling pose an imminent danger.[4] Feldman went so far as to say the administration's motives seemed to be "driven by political or social agendas on all sides."[5]

Unfazed, the Interior Department issued a second challenge, but that appeal was rebuffed on July 8 by a three-judge panel assigned to the case in the Fifth District Court of Appeals. The panel stated it was open to a further hearing on the merits of the appeal in September. However, the government wasn't interested in waiting for that. Instead, Interior Secretary Ken Salazar painted the getaway car a different color and quickly sped away to pull his next caper. On July 12, Salazar issued, not a moratorium, but a "suspension" on all floating-type rigs, like the one used by BP, in any depth of the Gulf's waters, through November 30. About thirty-six rigs were instantly impacted.[6]

What followed was an unpublicized meltdown of the Gulf oil industry. Getting a permit from the feds to drill anywhere became nearly impossible. As for the thirty-plus floating rigs that were commanded to cease operations, the majority packed up for other waters. Each of these massive rigs (the larger structures are 800 feet wide, 150 high, and weigh some 70,000 tons) cost up to $600 million to build, and are constructed under strict debt-financing terms that require the owners to have long-term drilling contracts secured in advance. If those contracts are not generating revenue to pay back their loans, the rig owners must wind up operations and go to where the money is.

Team Obama deviously sucker-punched Big Oil square in the pocketbook.

Eric Smith, a business professor at Tulane University, calculated that the Gulf moratorium and suspension caused a combined loss of 137,000 jobs, with the state of Louisiana forfeiting up to $400 million in oil tax receipts.[7] Professor Joseph Mason of Louisiana State University estimated the overall economic loss for the Gulf region—not because of damage from the BP spill, but because of Obama's decisions—at $3 billion.[8]

As the suspension drew toward its expiration, the administration had the audacity to make it appear as if they were graciously lifting the ban. Ken Salazar took the stage, bogusly boasting, "The policy position that we are articulating today is that we are open for business."[9]

However, "open for business" was yet another shrewd ruse. A company could drill, but first they had to locate a floating rig, and then jump through the myriad of brand-new rules and regulations contrived by Salazar for obtaining a new permit. Executives in the oil industry refer to Salazar's de facto moratorium as a "permitorium."[10] Six months after Salazar's supposed reboot, only four permits had been issued—all to projects that had been suspended a year earlier.[11]

Because of Obama's real, and de facto, moratoriums, the United States lost an estimated 360,000 barrels of oil production per day from the Gulf of Mexico in 2010 and 2011.[12] According to the Energy Information Administration's Annual Energy Outlook, released April 26, 2011, the combination of the new permitting rules and the lasting effects of the drilling moratoria will cause offshore oil production to continue lower than expected "throughout most of the projection period," which extends to 2035.[13]

Meanwhile, in the Arctic, where drilling applications were essentially suspended indefinitely, Shell finally conceded defeat. Speaking on a conference call with investors and media in February 2011, Shell CEO Peter Voser urged the Department of Interior to stop the stonewalling and speed up the permitting process. "There will be no drilling offshore Alaska in 2011," he said. "We need urgent and timely action on permitting to go ahead with the 2012 drilling program."[14]

If Mr. Obama really wanted to make good on not leaving our energy problems "for the next president," then he needed to quit playing games with the American people and take advantage of the plentiful oil reserves we have right here in the United States, oil fields that could keep us from purchasing one drop of foreign oil for *hundreds of years.*

But instead, left-wing radicals continue to utilize oil spills as effective tools to restrain an industry from providing us with a vital product. In the process, they are preventing free enterprise from creating permanent jobs and wealth, and keeping our country dependent on the resources of rogue nations.

SPILLMONGERS

Just as Marx valued using pollution as a blunt object to beat capitalism over the head, his current apparatchiks are no different. In their demonization of Big Oil, there is no finer weapon in their backpack than an oil spill.

As we discussed in chapter 2, the 1969 oil discharge off the coast of Santa Barbara is still commemorated each Earth Day by activists in California. Their vigilance has paid off quite well: West Coast oil drilling has never expanded beyond a small sector offshore from Southern California, despite the fact that there are plentiful, easy-to-access reserves up and down the entire coast. In 1978, the left-coast anti-oil crowd scored their greatest victory against Big Oil, and the American people, when the Outer Continental Shelf Lands Act of 1953 was amended by Congress. What's amazing is the 1953 Act was originally passed to ensure that the massive oil reserves off our coasts would be *available* for harvest. The Act declares, "The outer Continental Shelf is a vital national resource reserve held by the Federal Government for the public, which should be made available for expeditious and orderly development."[15]

The amendment of 1978 turned the original law on its head and gave the government power to actually cancel any offshore drilling lease or permit if it was even *thought* there could be "a substantial probability of significant impact on or damage to the coastal, marine, or human environment."[16]

This policy opened up the junk-science barn door. Now an anti-drill president could employ a like-minded biologist to produce a fraudulent study illustrating how a single species might be impacted if Big Oil dipped its wick off the West Coast. Instantly drilling could be scratched from the menu. But even the 1978 amendment wasn't enough for the eco-socialists; they desired absolute assurance that future drilling will never, ever, occur on the Shelf.

Their most recent attempt for such a law occurred in January 2011, with the Gulf spill still fresh in the public's mind. Ultraliberal Democratic senators Barbara Boxer (CA), Maria Cantwell (WA), Dianne Feinstein (CA), Jeff Merkley (OR), Patty Murray (WA), and Ron Wyden (OR) introduced a further amendment to the Outer Continental Shelf Lands Act. Under the plot, labeled the West Coast Protection Act, offshore drilling would be absolutely forever prohibited on the outer continental shelf of California, Oregon and Washington State.

In announcing the scheme Senator Boxer claimed, "We cannot afford to put California's coastal economy at risk by drilling offshore... The bill would help prevent another economic and environmental disaster like the Deepwater Horizon oil spill along the Gulf Coast last year."[17]

San Francisco Bay Area progressive Congressman and Nancy Pelosi surrogate John Garamendi introduced a similar bill in the House of Representatives. Thankfully, given the current makeup of Congress, these bills won't be passing anytime soon. Nevertheless, this illustrates why elections matter. Left unchecked, leftist politicians will always take advantage of a crisis, real or imagined, to attain their green goals. Such is the case regarding another oil misadventure from 1989—the *Exxon Valdez* oil spill. And it had nothing to do with drilling.

The *Exxon Valdez* was an enormous 980-foot supertanker loaded with 55 million gallons of oil (1.3 million barrels). The oil had been drilled in Prudhoe Bay, located in the far northern reaches of Alaska, near the Alaskan National Wildlife Refuge, and had been transported eight hundred miles through the Trans Alaska Pipeline to the tidewater port in Valdez. The ship departed for Long Beach, California, shortly after 9 p.m. on March 23, and soon discovered the outbound shipping lane was obstructed with small icebergs. Captain Joseph Hazelwood received permission from the Coast Guard to alter his course to the inbound lane. What happened next has been the subject of many books and numerous lengthy articles. Whether it was faulty radar, an understaffed crew, or an intoxicated captain, we may never be sure. However, we do know that shortly after midnight, the *Exxon Valdez* ran aground on the Bligh Reef, and at least 11 million gallons of oil (262,000 barrels) were spilled into Prince William Sound.[18] Prince William Sound's remote location, accessible only by helicopter, plane, and boat, made existing government and industry response efforts very difficult. The oil eventually washed up onto thirteen hundred miles of rocky coastline, enabling the enemies of Big Oil to have a field day with the dramatic photographs of harmed wildlife.

Booms were spread out on the open water and vacuums were brought in to suck up the oil. However, because the Sound is a calm channel, the lack of agitating waves pounding the coast prevented the oil from naturally becoming displaced, so the muck literally clung to shore. If the same spill had occurred along the rugged south coast of Oregon, where tumultuous waves continually crash into a daunting landmass, it would soon dissipate without human aid. Not so in Prince William Sound.

Also, keep in mind the dark crude, though messy, is 100 percent organic (in fact, the word *organic* refers to matter such as oil), and over time it is biodegradable. However, because of the striking photos, theatrical news reports, and inflamed rhetoric from the environmentalists, pressure was placed on clean-up crews to not patiently wait for nature to move into action, but instead to employ immediate manpower to clean the rocky shore. The decision proved disastrous. High-pressure hoses spraying hot water were brought in to blast the shoreline, and in the process, the pressurized water displaced and destroyed the coastal microbial organism populations (plankton, bacteria, and fungi) which are naturally capable of facilitating the biodegradation process. This exacerbated the cleanup efforts.

By the way, the natural biodegradation process is the reason remnants of the BP Gulf oil spill vanished more quickly than most media pundits, liberal politicians, and environmentalists seemed to prefer.

Exxon spent an estimated $2 billion cleaning up the spill in Prince William Sound, and another $1 billion to settle related civil and criminal charges. After many years of litigation, in 2008 the Supreme Court forced Exxon to pay $507.5 million in punitive damages.

In the aftermath of the accident, corrective measures were employed via an executive order issued from the Alaska governor's office, requiring two tugboats to escort every loaded tanker from *Valdez* out through the Sound to the open waters. Eventually the tugboats were replaced with a 210-foot escort response ship. Additionally, unlike the *Exxon Valdez*, tankers serving the region today are double hulled, lowering the risk of future problems.

The lesson to be learned from this anomalous accident is not that we should ban drilling in Alaska, but that we must learn from errors, make corrections, and move on. Nonetheless, *Exxon Valdez* lives as a green icon for prohibiting new drilling anywhere in Alaskan waters—that includes in an oil-rich region adjacent to the remote Aleutian Islands, called Bristol Bay. It wasn't until 2007, when the Bush administration lifted a long-standing moratorium on drilling in Bristol Bay, that oil companies were granted access to begin plans to harvest the 230 million barrels of oil, and 6.8 trillion cubic feet of natural gas, accessible in that isolated corner of the world.

But once again, in keeping with his mission to promote the green agenda, on March 31, 2010, President Obama stepped in to quietly *reverse* the Bush decision to tap Bristol Bay. It's yet another move made by this administration to keep us dependent on oil produced by a gaggle of nations with a lengthy history of not having America's best interests in mind.

AMERICANS HELD HOSTAGE

Since 1960, we've elected federal leaders who have allowed us to be held hostage by a small cabal callous to Life, Liberty, and the Pursuit of Happiness. The countries in this clan have proven to be interested in lining their own pockets with American dollars, while forcing us to tolerate their generally evil ways.

They are the nations of OPEC.

The Organization of the Petroleum Exporting Countries—OPEC—was founded in Baghdad, Iraq, in 1960 by five countries: Saudi Arabia, Venezuela, the Islamic Republic of Iran, Iraq, and Kuwait. These nations are OPEC's "Founder Members" and keepers of the world's largest ready-to-drill-now oil reserves. The quintet has masterfully arranged a world system in which they are the primary arbiters of the price per barrel of oil, because they control the majority of the supply.

Saudi Arabia's ruling royal family possesses the largest "proved discovered" oil reserves on the planet (I'll explain the difference between the types of reserves in a moment). It's estimated that Saudi Arabia holds 25 percent of the world's oil. If you want to know why President George W. Bush famously held the king of Saudi Arabia's hand (a Sharia-loving monarch, who treats women like chattel, hates Jews, and will not tolerate a single Bible brought within his borders) during a visit to his Texas ranch in 2005, it's because our two countries are codependents.

Venezuela owns the second-largest proved discovered oil reserves. It's a land run by the communist kook Hugo Chavez, the dictator whom

President Obama initially greeted with a bro shake (if you'll recall, Obama bowed upon his first encounter with the Saudi king).

Iran has the third-largest proved discovered reserves. It, too, is a nation run by extremists; in this case Jihad-loving Muslim clerics and a puppet president who denies the Holocaust.

Iraq is home to the fourth-largest proved discovered reserves. Iraq's former dictator, Saddam Hussein, was the most maniacal of the OPEC crew, and the other founding members were likely pleased to see him eliminated.

Kuwait is keeper of the planet's fifth-biggest proved discovered oil deposits. Kuwait's ruling family has been quite generous with their oil revenue, taming the subjects of their small country by turning it into a highly developed welfare state, shrouded in a quasi-free-market economy.

These five Founding Members rule OPEC with an iron fist. Their bylaws allow other nations to join their club as "Full Members" as long as the state in question has "a substantial net export of crude petroleum, [and] fundamentally similar interests to those of Member Countries."[19]

One can only wonder what "similar interests" refers to? Given that the Full Members are Algeria, Angola, Ecuador, Nigeria, Qatar, and United Arab Emirates, it would appear as if "similar interests" include assorted shades of tyranny, as all Full Member states are highly influenced by either Islamic Sharia law or varying degrees of Marxism. This is also very important: all OPEC countries have state-run oil companies, and it's becoming increasingly difficult for for-profit corporations to do business with them. For example, in January 2011, Ecuador, who for decades allowed foreign oil companies to negotiate contracts to lease oil fields in their country, imposed new conditions making it unprofitable for Big Oil to do business there. "To the companies that do not accept the new conditions, we'll say to them good-bye and good luck," expounded Ecuador president Correa.[20]

In 2007, Venezuela's Chavez took a more hair-raising stand against Big Oil by literally taking over the last remaining oil production sites that were under foreign ownership in his country.

And the truth must be told in Iraq. During Saddam Hussein's three-decade rule, all of the oil beneath his soil belonged to him, and his cruel sons—and they weren't known for sharing the wealth with their people. Since the 2003 invasion and liberation of Iraq, there are those who claim that the primary motivation for President Bush and Vice President Dick Cheney was to gain control of the world's third-largest oil reserves. However, those assertions were dealt a blow in 2009, during an auction of Iraqi oil contracts. It was the biggest sale of oil field leases in Iraq's history, and amazingly, not a single U.S. company secured a deal in the auction. In fact, the most lucrative multibillion-dollar contracts went to two countries that bitterly opposed our campaign in Iraq—Russia and China. Iraq claimed up front that no political considerations would be made for oil leases; a damned shame given that a lot of American blood was shed to free the Iraqi people of the Hussein family.

Together, the eleven members of the OPEC cartel own 79 percent of the world's proved discovered reserves and 44 percent of the world's oil production.[21] Certainly one of their principal goals is determining the best means for safeguarding the organization's financial interests, individually and collectively. It's a scheme that's worked brilliantly, particularly for the royal families who rule Saudi Arabia, Kuwait, Qatar, and United Arab Emirates, as well as for Hugo Chavez, who can fund his brand of Marxism with profits from his state-run oil empire.

WHO'S REALLY MAKING ALL THE MONEY?

OPEC's ability to manipulate supply and set the price of oil has worked masterfully to increase their bottom line. In 2002, crude oil was selling for $25 a barrel. Five years later, it had risen to $100, allowing profits to rise exponentially. But of course, the jump in per-barrel price was not just good for OPEC. As the price of the commodity rose, every player in the industry had an opportunity to prosper, including the shareholders of commercial American-based oil corporations, which include millions of everyday folks with IRAs, 401(k)s, and pensions. However, at the pump we were all feeling the pain—and so were the top U.S. oil executives.

In early 2008, with the per-barrel price nearing $140, and gasoline hovering near $4 per gallon, leaders from the top oil firms were frog-marched onto Capitol Hill to endure socialist-sounding lectures from Democratic members of Congress. The oil execs were being charged with making too much money off the backs of the American people. After reading the first half of this book, you know what was really occurring.

Marxist ideologues were attacking businesspeople who were producing a vital commodity garnered from a natural resource, and making a profit in the process.

That's an unacceptable combination to any Marxist ideologue.

According to congressional research procured for the mock trial, ExxonMobil profited the greatest with their 2007 net revenue hitting a record $404.5 billion, and a net income of $40.6 billion—the most ever by an American company. In fact, the *New York Times* added, Exxon's record revenue "exceeded the gross domestic product of all but the 24 richest countries in the world."[22]

What the *Times* failed to mention was that Exxon's profit margin (return on sales), as calculated by Congress, was a modest 10 percent. Shell was next on the big-money list, with $335.7 billion in sales, and $27.5 billion in profits (a 7.1 percent profit margin). BP was third, with $291.4 billion in revenue, and $17.2 billion in profit (and a lackluster 5.9 percent margin).[23]

Senator Diane Feinstein, whose personal net worth is estimated to be as high as $110 million,[24] scolded the businessmen, "You rack up record profits, record profits for any corporation in the United States of America, quarter after quarter after quarter, and apparently have no ethical compass about the price of gasoline."[25] Senator Dick Durbin of Illinois added, "Does it trouble any of you when you see what you are doing to us, the profits that you are taking, the costs that you are imposing on working families, small businesses, truckers, farmers?"[26]

Senator Barack Obama was on the presidential campaign trail during the hearings, but took the opportunity to pile on, stating that ExxonMobil was guilty of "making money hand over fist, making out like bandits."[27]

So, if Exxon's $40 billion in profit was akin to robbery, I wonder what Obama, Durbin, and Feinstein thought about the $200 *billion* in profit Saudi Arabia's private oil firm, Aramco, made that same year? That's almost twice as much money as the combined profits of the top nine oil companies Congress was investigating.[28]

At the time, *Forbes* magazine said Saudi Aramco was "easily the most profitable company on the planet."[29]

Why is Aramco so profitable? Because oil practically oozes from the earth's surface in Saudi Arabia. Aramco's CEO told reporters that soaking up a barrel of crude from their reserves costs a mere $2.[30] U.S.-based companies have to work a lot harder to pull a barrel out from beneath American property, thus the average cost of our land-based oil runs about $25 per barrel, while out in the Gulf of Mexico, it's closer to $70.

And when it comes to those "bandits" Obama bad-mouthed, the real masked man is the government.

Though Exxon did earn $40 billion in 2007, it also paid $30 billion that year in taxes.[31] It was the largest tax check written to the Treasury that year. In 2008, Exxon's effective tax rate was 49 percent.[32]

Interestingly, in the twenty-five years leading up to Exxon's record revenues, the U.S. oil industry had collectively earned profits of $643 billion (adjusted for inflation). In contrast, during that same period, federal and state governments hauled in $1.34 trillion in taxes from gasoline sales alone.[33]

And don't forget the amount of money you and I pay to the government in taxes at the pump. Every gallon of gas we put in the tank includes a federal excise tax of 18.4 cents, plus a state excise tax averaging about 24 cents per gallon. In addition, eleven states levy other taxes and fees, adding to the cost of filling up.

HOW MUCH OIL?

In a televised address on April 18, 1977, President Jimmy Carter delivered a chilling prediction: "Unless profound changes are made to lower oil consumption, we now believe that early in the 1980s the world will be demanding more oil that [*sic*] it can produce... Within ten years we would not be able to import enough oil from any country, at any acceptable price."[34]

It was just one of many preposterous things the former peanut farmer said. First of all, to imply we will run out of oil dismisses laws of economy. Something of value never runs out—its price just rises higher. More important, Carter revealed that he held a subscription to a failed theory known as "peak oil." Based on junk science and held tightly by environmentalists, peak oil claims the supply of available oil has peaked, and henceforth humans are recklessly sucking up the last remaining supplies. Peak proponents usually declare we'll run out of oil in twenty-five to thirty years; back in Carter's day it was a decade.

Now contrast Carter's quote to this one: "We are looking at more than four and a half trillion barrels of potentially recoverable oil. That number translates into 140 years of oil at current rates of consumption, or to put it another way, the world has only consumed about 18 percent of its conventional oil potential. That fact alone should discredit the argument that peak oil is imminent and put our minds at ease concerning future petrol supplies."[35]

Four and a half trillion barrels of oil.

The remark was made by Abdallah Jum'ah, CEO of Aramco, during an OPEC meeting in Vienna, Austria, September 13, 2006. And Jum'ah didn't just randomly pull that figure out of his turban. It's an amount determined through a summation of OPEC's "proved undeveloped" oil reserves.

Now a quick lesson on oil reserves. There are two types—"proved developed" and "proved undeveloped." *Proved developed* reserves are those quantities of oil that, by analysis of geological and engineering data, are estimated with a high degree of confidence to be commercially recoverable, under present economic conditions, from known reserves. Think of these reserves as low-hanging, easy-to-pluck fruit on a tree. Proved undeveloped reserves are reserves that are expected, with a high degree of certainty, to be recovered from future wells and facilities, as technology and economic conditions allow. In other words, the proved undeveloped reserves are like trees loaded with fruit, but the fruit is more challenging to reach because it's fastened high on the tree.

Distinguishing these differences in reserves is particularly important to the investor or accountant who is determining the financial feasibility of a proposed drilling project. He or she is concerned with how easy, or difficult, the oil will be to extract from the ground, how much refining will be required before the oil gets to market, and how far the reserve is from its marketplace. Reserves that fall below a standard index of affordability are not developed because they would not be positive financial ventures. Thus, proved undeveloped reserves are not listed with proved developed reserves. However, as economic changes occur, technology improves, or supplies tighten, the proved undeveloped reserves switch across the ledger.

For example, in 1944, Middle Eastern proved discovered reserves were estimated at 15 billion barrels. By 1975, those same fields had produced 42 billion barrels, and there were said to be 75 billion proved developed barrels remaining. By 1984, that number of proved developed barrels increased to 398 billion. As 1990 approached, proved developed Middle Eastern reserves were said to be 654 billion barrels.[36] Today the number is closer to 800 billion.

Currently, global daily oil usage is estimated to be 85 million barrels per day (31 billion barrels per year). The peak oil simpletons look at these numbers and go into a Chicken Little imitation, squawking, "Only 25 years left. Gotta stop driving cars now!"

However, with OPEC laying claim to four and a half trillion barrels—the peak oil theory quickly disintegrates. There's enough oil for the nations of the world to grovel at the feet of a cartel filled with radical Muslims, socialists, and communists for the next 140 years.

But it need not be. The United States has a comparable cache of oil under its own soil too—and the OPEC honchos know it. And to prevent our tapping this colossal supply, they play us like a cheap sitar.

Allow me to illustrate, by developing a simple business analogy.

You own a coffee shop and purchase coffee beans from me. I have more coffee bean trees than anyone and can grow the beans really cheap. Your customers love coffee—in fact, they're practically addicted to it. I've been supplying you with great beans for as long as anyone can recall, and your coffee is very good. You make money, I make money, and your patrons are happy.

However, I'm actually a very greedy guy, and a slick operator who craves giant mansions, fast cars, fancy clothes, rich food, and a big bank account. So, occasionally I decide to raise the price of the beans you buy, to increase my profits, knowing you really don't have much of a choice—you will never grow beans as cheap as I'm able to.

It's quite a racket. Whenever I want more money, I just raise the price of your beloved beans, and you in turn raise the price of your coffee. Your faithful customers temporarily complain, but come right back for more. Eventually everyone seems to get used to the higher-priced cup o' joe.

"Bunch of suckers," I say to myself. You see, I'm getting a kick out of this arrangement, especially since I really don't like you, and only tolerate you for one reason—your money.

One day an agronomist friend of yours shares a plan with you on how to start growing *your own* coffee plants. You discover your climate is actually perfect for it—all you need is some capital to buy or lease some land, plant the trees, and wait about four years for them to mature. If you do it properly, within a couple decades you would never have to buy a *single bean* from me again.

Your customers get wind of this plan and are excited; after all, the cost of their coffee could be greatly reduced, jobs would be created in your coffee fields, and they would never have to deal with my company again. They start a campaign, urging you to "plant, baby, plant!"

I quickly find out about your fanciful ideas and act immediately by slashing the cost of my beans in half. It's really not that big of a deal for me. Heck, I'm still clearing lots of profit.

In the meantime, I'm intrigued to see your response to my 50 percent price cut. You've reduced the price of a cup 25 percent, pocketing the other 25 percent for yourself. You say the extra profit will be used to improve your facilities.

I'm impressed. You're a smart businessperson after all, and your customers are content too. Now they're purchasing a couple more cups a week.

Oh, and of course, I'm especially thrilled that you've dropped those silly notions to *plant, baby, plant.*

This, friends, is exactly what OPEC does to America every time we get serious about domestic drilling. Because their member countries possess 79 percent of the proved developed reserves and hold a combined 44 percent of all production, OPEC is able to manipulate the market by setting caps on their share of the world's oil production. By increasing supply, they cut the price of crude and watch us cool down like programmed drones.

U.S. oil production totals 5.5 million barrels per day.[37] But here's the imbalance: we import 2.1 million barrels daily from Canada, 1.2 million from Mexico, and a total of 3.9 million from the OPEC countries of Saudi Arabia, Nigeria, Venezuela, Iraq, and Algeria.[38] However, instead of allowing us to utilize our own resources and reduce what we are importing, particularly from OPEC, the Obama administration has chosen to subsidize our foreign importers.

In September 2010, the U.S. Export-Import Bank guaranteed $1 billion in loans to Pemex, the Mexican state oil company, to bolster the company's oil drilling in the Gulf of Mexico.[39] While Obama is working overtime to keep Big Oil out of the Gulf, he enabled state-owned oil to drill, baby, drill.

The Export-Import Bank is the official export credit agency of the United States federal government. It was established in 1934 by President Roosevelt via executive order, and was made an independent agency in the executive branch by Congress in 1945, to finance and insure foreign purchases of American goods for customers unable or unwilling to accept credit risk. The agency also loaned more than $1 billion to Mexico's state oil company in 2009. Because the bank is an independent agency, the deal was not subject to congressional approval.

Even more appalling, in 2009, President Obama cleared the way for the Export-Import bank to lend billions of dollars to Brazil's state-owned oil company, Petrobras, to finance exploration of the huge offshore discovery in Brazil's Tupi oil field in the Santos Basin near Rio de Janeiro. Brazil is our tenth leading oil importer.

Our dependence on, and aid to, these foreign entities, is pathetic.

Team Obama is clearly following the Marxist view that state-owned oil is superior to private companies who exact a profit and create wealth from the natural resource they harvest.

Here's what you need to know about America's oil supply. Proved developed reserves are stated to be 22 billion barrels.[40] Most of it is concentrated in Alaska, Texas, U.S. waters in the Gulf, California, and North Dakota. Annual U.S. consumption is 7 billion barrels.[41]

If we were to rely exclusively on our own domestic proved developed reserves, we'd be back to the foreign oil trough in just a few years. However, it's our proved *undeveloped* reserves that cause OPEC to get fidgety, and that's the resource on which we should be planning our future. The fact is, there is enough domestic crude to keep the U.S. flowing in oil for generations—in fact, we'd never have to import a drop of oil from any country, because it's estimated that we have as many trillions of barrels as OPEC does.

OPEC knows this very well, and so do the radical environmental organizations, eco-extremist politicians, and the Obama administration. Together they are engaged in an all-out battle to prevent us from harnessing our own, liberty enhancing, black gold.

NINE

OIL INDEPENDENCE

OIL HAS BEEN VERY GOOD for the people of Alaska. The folks of "the last frontier" have prospered because of the oil industry. In each year since 2001, at least 80 percent of unrestricted state revenue has come from petroleum.

According to the Institute of Social and Economic Research at the University of Alaska, the average Alaskan family of four enjoys an estimated yearly value of $22,000 in tax relief, state-supplied dividends, and enhanced public services.[1] That same research concludes that there are 127,000 jobs directly attached to the state's oil industry, and half of all Alaskan jobs are in some way associated with oil development.

Alaska's proved developed reserves are stated to be 3.5 billion barrels. However, it's been estimated, by the Department of Interior, that in a portion of extreme northern reaches of the state, there could be as much as 29 billion barrels of oil. The problem is the reserve is located in the Alaskan National Wildlife Refuge (ANWR) and a well-organized, highly determined group of environmentalists has been fighting for years to keep Big Oil out.

ANWR (pronounced "ann-wahr") lies in the northeast corner of the Union's largest state, just above the rim of the Arctic Circle, thirteen hundred miles south of the North Pole. In some locations, crude oil is so plentiful that decades before Alaska became a state in 1959, explorers reported oil literally seeping up from beneath the ground.

However, in the 1930s, a vocal socialist who worked as a federal bureaucrat in the Departments of Agriculture and Interior began a campaign to keep any form of capitalism from treading on the great Alaskan territory. The activist's name was Robert "Bob" Marshall. Born in 1901 to a life of great privilege in New York City, Marshall was the son of one of the nation's most prominent liberal constitutional lawyers, Louis Marshall, who argued more cases before the Supreme Court in his time than anyone. Bob Marshall attended the New York State College of Forestry at Syracuse, earned a master's in forestry from Harvard, and eventually completed a PhD in plant physiology at Johns Hopkins. It was while pursing his doctorate that Marshall acknowledged becoming a full-fledged socialist. "I wish very sincerely that Socialism would be put into effect right away and the profit system eliminated," he boasted.[2]

In 1933, Marshall published *The People's Forests*, which proposed a radical solution to save the forests of America from drilling, damming, logging, and even dirt trails and roads. Marshall advocated for the public acquisition and administration of all forestland in the United States. In keeping with the theme of similar dictates written by Lenin, such as *On Land* and *On Forests*, the last words of *The People's Forests* succinctly articulate Marshall's seditious spirit: "The time has come when we must discard the unsocial view that our woods are the lumbermen's and substitute the broader ideal that every acre of woodland in the country is rightly a part of the people's forests."

Marshall's work proved highly influential. He is credited with shaping the U.S. Forest Service's policy on wilderness designation and management, and personally underwrote a new government map of regions of America that should be considered "roadless areas" (*roadless* implies that no roads, trails, or paths of any kind be allowed in a region). And regarding Alaska, he vigorously proposed creating federal laws to prevent human intrusion into millions of acres of the state's northland.

To see his mission better realized, in 1935 Marshall used family money to help found the Wilderness Society.

We must pause for a moment to detail the Wilderness Society, as the fanatical, uncompromising organization will be discussed often throughout the balance of this book.

The Wilderness Society (WS) has never seen an oil rig, lumberman, dam, mine, hunter, or off-road vehicle it didn't despise. WS takes credit on their website for having "been at the forefront of nearly every major public lands victory over the past 75 years." Indeed, WS has permanently lobbied to institute laws to prevent natural resources from being harvested within 575 separate regions of the U.S. And they've been extremely successful assisting the government in declaring nearly 110 million acres of federal land, in forty-four states, "wilderness." WS defines wilderness as an area that prohibits "mechanical devices, human structures, resource extraction and all activities harmful to the land."[3] Given that the U.S. owns nearly 700 million acres of public land, the Society has effectively aided in shutting down about 16 percent of it. And as you'll see, their plans are to take it all.

WS marches in lockstep with the demands of the United Nations. They spend nearly $25 million each year in the U.S. to lead "on-the-ground campaigns that mobilize the public and local, state and national decision-makers" to achieve their goals.[4] Proving their influence inside the Washington, D.C., beltway, during the Clinton-Gore administration, the White House Office of Management and Budget's deputy director was Alice Rivlin, who formerly served as chair of the Wilderness Society's Governing Council.

Another cofounder of the WS was Marshall's close friend Aldo Leopold, who is revered as a saint in today's green movement. In 1922, Aldo called for "radical action," demanding that "either the state or national government put all land under inspection."[5] Regarding roadless areas, eco-zealots love to quote Leopold, who said that society should not promote the enjoyment of the great outdoors through "building roads into lovely country, but of building receptivity into the still unlovely human mind."[6]

Bob Marshall mysteriously died in 1939, at the age of thirty-nine, but the Wilderness Society would carry his desires forward. Olaus Murie, one of the organization's first members and eventually its president, and his wife, Mardy, pushed for the establishment of an Arctic wilderness area in Alaska. After considerable lobbying, in 1960, the Eisenhower administration acquiesced to their demands and set aside 8.9 million acres to form the core of what is today ANWR. According to the Wilderness Society website, Mardy Murie later recalled how the news of the establishment of

ANWR moved her then elderly husband to tears. She said it was one of only two times in a forty-year marriage that she saw Olaus cry.[7]

If old Olaus shed a tear when Eisenhower designated ANWR, he would have bawled like John Boehner had he lived to witness President Jimmy Carter further expand the wilderness area by signing the Alaska National Interest Lands Conservation Act in 1980.

ANWR now totals 19 million acres.

Of all their exploits, ANWR is the Wilderness Society's greatest trophy. In laying claim to this vast corner of Alaska, the Society has insidiously cheated Americans out of a spectacular supply of oil, derived from just a sliver of the vast region—a sliver that is virtually uninhabitable to human life.

NO-MAN'S-LAND

When environmentalists trot out their propaganda photos of ANWR, designed to dissuade those interested in drilling, their subjects include cute, furry foxes, a mother bear and her cubs, inspirational mountain peaks, and glorious flowering meadows. What these disingenuous eco-hucksters don't tell you is these photos were not taken in the region of ANWR where drilling would occur.

The northern rim of ANWR is known as the *coastal plain*, and the oil-producing area of that region is known as "10-02," which refers to Section 10-02 of the 1980 Lands Conservation Act. Section 10-02 set aside 1.5 million acres of the Arctic Coastal Plain specifically for, not wilderness or refuge, but "oil and gas exploration," because of its well-known potential for large deposits of oil. Further, Section 10-02 specifies that any land used for exploration would be limited to a footprint size of 2,000 acres.[8] This means that within the 19.6 million acres of ANWR, only a fraction of a percent of the total land may be used for drilling.

Authentic photos of 10-02 would reveal it's a no-man's-land, bordered on the north by the stormy waters of the Beaufort Sea, on the east by ANWR's original wilderness area and the U.S.-Canadian border, and on the west by the Canning River and ANWR's outer border. It is one and a half million acres of completely flat, barren land. Nine months of the year, the vast expanse is covered with snow and ice, and virtually void of most animal life. Three of those winter months are draped in round-the-clock darkness. As the snow melts in late spring, the coastal plain is dotted with thousands of temporary lakes and bogs. Summer skirts by in a brief six weeks. That month and a half is noted for mobs of large mosquitoes, and swarms of flies are drawn to the extensive mounds of manure from wandering caribou.

10-02 was made to drill, baby, drill.

Yet despite the act created in 1980 to drill, year after year liberals in Congress have continually bowed to the wishes of the Wilderness Society, keeping even the slightest hint of exploration from occurring in ANWR. This has become an annual slap in the face to the Alaskan people, who continually make clear their desire to see the 2,000-acre parcel developed. Every sitting Alaska state legislature has passed near-unanimous pro-ANWR resolutions since 1980. This result flies directly in the face of the reception the issue gets in Washington, D.C., where Congress yearly deals with numerous ANWR bills, and has played partisan political football with the issue for more than thirty years. ANWR remains the single most legislated energy issue in American history. Legislation to drill in ANWR has separately passed the House twelve times, and the Senate three times. However, only once have both bodies simultaneously agreed to legislation that was sent to the White House. That occurred in 1995, and their bill was dead on arrival, as it was quickly vetoed by Al Gore's boss, President Clinton.

Since early 2006, there has not been a floor vote on Capitol Hill regarding tapping this essential resource. In fact, the issue only managed a single committee amendment vote in 2009.

This is another example of why elections matter. Reasonable representatives need to be elected into congress so that the ghost of Bob Marshall and the wishes of the Wilderness Society, are forever kicked into the dungeon of dangerous designs.

OIL SHALE: AMERICA'S GREATEST FOSSIL RESOURCE

While 29 billion barrels of Alaskan crude would be a nice addition to America's resource portfolio, there's another staggering cache of oil that would make the CEO of Saudi Aramco soil his tunic if the U.S. ever decided to tap it. It's known as the "Green River Formation in the Uinta Basin" of the Rocky Mountains, and the oil is found within a fine-grained sedimentary rock known as *oil shale*. Oil shale contains kerogen, which, when heated to extreme temperatures, yields oil.

The vast extent of American oil shale resources has been known for a century. In 1912, President Woodrow Wilson, by executive order, established the Naval Petroleum and Oil Shale Reserves. This office, now within the Department of Energy, has overseen the U.S. strategic oil shale interests since that time. Though this federal office has tried to stimulate several commercial attempts to produce oil from shale, they have failed primarily because of the historically lower cost of petroleum oil with which it competes. Oil shale is more costly to "drill"; however, with the current high price of traditional crude, the economic viability of oil shale looks to have arrived, and it could keep us from ever having to kiss the ring of another member of OPEC ever again.

According to a 2011 news release from the Bureau of Land Management (BLM), there are likely 2.6 trillion barrels of oil present in the Green River region, with 1.5 trillion of those barrels proved and ready for the picking:

> The United States holds more than 50 percent of the world's known oil shale resource, most of which is contained in deposits found in Colorado, Utah and Wyoming. These oil shale deposits are found within a total area of approximately 16,000 square miles between the three states, with about 72 percent of [*sic*] located on federal lands. The total reserve contained within these deposits is estimated to be 2.6 trillion barrels of in-place oil, and 1.5 trillion barrels of recoverable oil.[9]

One and a half trillion barrels of "recoverable oil" (recoverable oil is as close to proved discovered as you'll ever get) would be enough to completely run America at the current pace for the next fifty years. The "2.6 trillion barrels of in-place oil" is akin to proved undeveloped. Once we harvest these resources, we will be completely oil-independent well into the next century.

This phenomenal reserve is located on a sixteen-thousand-square mile plot shared by Colorado, Wyoming, and Utah. The problem is, 72 percent of this property is on federal land, and the Wilderness Society is working feverishly to lock it up.

Late in 2008, as the Bush administration was winding up its final term, a decision was made to allow oil shale production to begin in the Green River Formation. The arrangement would have cleared the way for Shell Oil to gain access to a small tract of federal land and conduct a pre-commercial demonstration of shale technology, something in which Shell has invested millions over the years. Shell's project would have a minimal footprint, produce a modest one thousand barrels of oil a day, and require $200 million in up-front capital.

But as soon as the Wilderness Society got wind of this plan, they went berserk. Predictably, their first canard was, "The Bush administration pushed through last minute oil shale regulations in order to benefit their friends in industry."[10]

WS's second beef involved water; to properly release the oil from the shale, a considerable amount of water is needed. However, local governments, as well as the U.S. Department of Energy, have stated there is adequate water available from a variety of sources for the project to go forward. If the demonstration was found to be successful, there are ample locations in the region for reservoir creation, which, of course, the environmentalists would immediately assault with a hail of leftist lawyers.

The third complaint from the Wilderness Society was (no surprise here) global warming. Society activists claimed they're concerned about the process of heating shale to temperatures in excess of 600 degrees Fahrenheit, in order to release the oil. The energy needed to heat and process oil shale, they say, would increase the global warming emissions that change the climate.

In a letter of complaint to the U.S. Department of Interior, the Wilderness Society whined about "inaccurate estimates of water available in the Colorado River Basin to support a commercial oil shale industry and the BLM's utter disregard for the potential global warming impacts of pursuing oil shale without significant additional research."[11]

The "significant additional research" demanded by the Wilderness Society was granted in due time. Lawsuits were filed against the government by thirteen different eco-organizations (referred to by the media as "conservation groups") in attempts to stop the proposal. As soon as "W" left office, the Obama administration put the plan on ice, with Interior

Secretary Salazar stating, "We need to push forward aggressively with research, development and demonstration of oil shale technologies to see if we can find a safe and economically viable way to unlock these resources on a commercial scale."[12]

It was a deceitful remark. Translated properly, what Salazar really said was, "We need to push forward aggressively with research, development and demonstration of oil shale technologies to *assure there are* no *safe and economical ways to unlock these resources.*"

Proving my point, in a carefully worded announcement issued in February 2011, Salazar said the plan to harness shale oil was going to be shelved. The administration was caving to the demands of the Wilderness Society and others involved in the suit. Salazar claimed that new environmental rules and regulations were needed to ensure that air and water quality would not be impacted from such a project. The Associated Press covered it like this:

> DENVER—The Obama administration has proposed revising rules and regulations for oil shale development on public land in the Rockies, Interior Secretary Ken Salazar said Tuesday.
>
> Salazar made his announcement while government officials filed a proposed settlement of two lawsuits by 13 conservation groups over the Bush-era plan to open nearly 2 million acres of public land in Wyoming, Colorado and Utah to the commercial development of oil shale and tar sands.
>
> . . . The settlement proposes requiring environmental reviews of commercial development, including impacts on air and water quality and water supplies.[13]

And with that pronouncement, one of the most prolific oil reserves on the planet was effectively shut down. The oil, the jobs, the tax revenue, and the freedom associated with buying American were all tossed aside for the green agenda. Team Obama stood in the way of liberating the U.S. from a radical cartel, hell-bent on keeping America as its highly favored sultana.

SHALE'S COSTS AND BENEFITS

We really can't let the Obama White House have the last word on this monumental shale deposit.

In 2006, the Department of Energy estimated that the development of Green Valley could be economically sustainable—depending on the quality of the shale—for as little as $35 a barrel and as much as $54 a barrel.[14] In an independent study for the DOE, the Rand Corporation projects the price per barrel would have to be $70 to $95 in order to maintain viability.[15] Rand estimates that if a modest 25,000 barrels were produced each day, in twelve years the price of the oil produced from shale would drop to $35 to $48 per barrel.[16]

The disappointment is, Shell was willing to take the first step, even though the cost to construct a plant capable of processing 100,000 barrels per day could be as high as $10 billion.[17] But Obama and crew blocked their path, and doing so stopped the clock toward true oil independence.

According to Rand, Shell Oil would need six years to design, obtain permits, and construct the demonstration plant. Then, after considerable testing, if the decision were made to go forward with full-scale commercial operation, it would require another six to eight years to get the first-of-a-kind plant online. Once in production mode, it would be seven years before Shell would be able to generate 1 million barrels per day, twelve years to reach 2 million, and seventeen years to attain the goal of 3 million barrels daily. Such production would be spectacular given that at our peak in the early 1970s, the U.S. was producing 10 million barrels per day.

If Obama hadn't pulled the plug on this plan, it's possible that by 2035—assuming oil was selling at a mere $50 per barrel—the domestic oil shale industry could be generating $50 billion per year. Not only would shareholders prosper under this scenario; so would the estimated fifty thousand Americans directly employed by the plant operator, plus the additional indirect employment in the community at large, likely including another hundred thousand people.[18]

And the government wins too. Based on current government tax schedules, half of a $50 billion annual gross profit would be distributed in tax revenue between federal, state, and local governments.[19]

If the Bush administration plan for Green River had been allowed to stand, we'd be four years into a twelve- to fourteen-year strategy to ramp up shale production to a viable position. Similar setbacks have also occurred due to Obama's closure of the seawaters off Alaska, as well as the plan to drill off the coast of Virginia. Those endeavors will need time in order to produce. President Obama has purposefully kicked our opportunity for oil independence backward.

And then there are the jobs that this administration keeps telling us they want to create. The U.S. spent more than $800 billion to "stimulate" the economy, with the administration boasting that such a scheme would see unemployment peak at just under 8 percent in 2009, and help create up to 4 million jobs by the end of 2010.[20]

Yet unemployment shot through the roof. It was all a reckless lie.

Do you want real jobs—not temporary gigs paving a road, building a solar plant, or erecting a wind turbine? Of course you do.

Then let's drill, baby, drill!

In the most recent complete survey of those employed in the fossil fuel industry, compiled by the Bureau of Labor Statistics, there are millions of jobs to suit every skill set. The current total number of full-time employees *directly* employed by oil or natural gas firms totals 3.3 million.[21]

However, that total does not include all of the millions of other important jobs associated with the oil and gas industry, including work at gasoline stations and for utility companies using natural gas. In the U.S., there are approximately 116,000 gasoline stations, 95,000 of those having a convenience store as part of the business.[22] According to census figures, nearly 900,000 people are employed by these establishments alone.

In the natural gas industry, 90,000 are directly employed in distribution and 84,000 work at plants that turn natural gas into electricity; all told, the number of people working for fossil fuel–based utility companies totals nearly 600,000.[23] Then there are the untold figures representing the various businesses that serve these establishments with maintenance, supplies, uniforms, and janitorial services. Literally, many millions of people are employed by our utilization of fossil fuels.

JUST DO IT

As noted in the last chapter, according to the Energy Information Administration, over half of the crude oil imported into the U.S. in 2011 came from OPEC. Our largest trading partners by the barrel are Canada and Mexico, followed by OPEC's Saudi Arabia, Nigeria, Venezuela, Iraq, and Algeria.[24]

Most of us have no issues with importing from our friends to the north, but our relationship with the oligarchy south of our border is troubling, to say the least. Given that Mexico's is a state-run oil firm, one would think they could sow the money back into their own economy and lift their land out of desperate crime and poverty. For decades, Mexican leaders have been accused of using oil profits to fund political campaigns and line personal pockets.

And then there are those of us in the U.S. who are disgusted by Islamic Sharia law. The Islamic nations of OPEC abide by such principles, which demand that they exact 2.5 percent of their profits for Muslim charities. Saudi Arabia has utilized their offerings to construct the majority of mosques in America—if that doesn't concern you, then perhaps the next sentence will. Sharia charities include organizations beholden to violent jihad. In fact, our federal government has classified twenty-seven Islamic charities as terror organizations; most recently, the Holy Land Foundation in Dallas, Texas. Several of its executive members were convicted of financing terrorism in 2008, and had been charged with funneling $12 million to the terrorist group Hamas. It's estimated there are approximately $1 trillion of Sharia assets managed all over the world today, with the vast majority of that money coming from the sale of oil.[25]

By supporting OPEC we are directly funding our own demise.

Now we need some American-style heroism in Washington, D.C., to allow our nation to responsibly develop our own resources, and, thus, cut the ties with those who do not share our values.

In addition to drilling, we need to construct new refineries. America currently has roughly 150 refineries, down from over 300 in the seventies. Due to the incessant whining of environmentalists, we haven't built a major new refinery in the U.S. since 1976. In addition to creating the various fuels we utilize, refineries are also our nation's primary source of oil storage. More storage equates with more security.

Building an energy-independent America is *real* economic stimulus that will end the massive transfer of wealth, in which we ship off hundreds of billions of dollars each year to nations that have not our best interests in mind. Thus, when a riot breaks out in Egypt, war occurs in Libya, Iran sends warships into the Strait of Hormuz, or Hugo Chavez trips while doing the cha-cha, we won't have to suffer the consequential rise in the price of crude. Decreasing our dependence on foreign reserves will reduce the impact of world events on our domestic economy.

In the end, energy independence is about Life, Liberty and the Pursuit of Happiness.

Drill it.

TEN

NUKE IT

LIKE ALL DEVOUT ECO-SOCIALISTS, Barack Obama is not fond of nuclear energy.

During the 2008 presidential primaries, when he was still rough around the edges and often carelessly exposing his true inner feelings, candidate Obama leaked his nuclear sentiments before a New Hampshire newspaper's editorial board. Asked about his stance on nuclear power, Senator Obama sarcastically replied, "I don't think there is anything we inevitably dislike about nuclear power; we just dislike the fact that it might blow up, and irradiate us, and kill us! That's the problem."[1]

The fawning media clucks chuckled.

Despite producing a carbon-free source of electricity, nuclear energy is adamantly opposed by environmentalists. They claim their primary disapproval is a nuclear accident that would spread deadly radiation. Like programmed drones, they immediately cite Three Mile Island, Chernobyl, and more recently, the Fukushima reactor in Japan, to provide ad hominem examples of nuclear power gone mad.

When rational individuals present arguments illustrating how such accidents have been overhyped, the environmentalist response focuses on the radioactive waste generated at the various nuclear plants.

That, too, is an easy argument to win, and it's where we'll begin.

There are 104 nuclear power facilities in the United States, with the oldest in operation since the 1960s. All of the nuclear waste ever generated within these plants is stored at 126 different sites located around the nation. The *total* amount of spent uranium (the primary waste material) amounts to sixty-five thousand tons.

Determining it would be best to safeguard these nuclear byproducts in one location, President George H. W. Bush signed the Nuclear Waste Policy Act Amendment in 1987, which, at the time, settled the waste issue by codifying Yucca Mountain in Nevada as the nation's first permanent nuclear waste repository. The deal gave Nevadans the opportunity to weigh in on the matter, but also allowed Congress veto power in the event the Silver State should reject the plan (even though Yucca Mountain is located on federal land).

Yucca Mountain is a rugged, barren, no-man's-land about a hundred miles northwest of Las Vegas. If you placed photos of Tora Bora, Afghanistan, next to those taken at Yucca Mountain, you wouldn't notice much difference. Yucca Mountain is an ideal location to secure nuclear waste.

After years of debate and billions spent procuring feasibility studies, in April 2002, Nevada governor Kenny Guinn gave the repository a thumbs-down. Since the governor's decision could only be overridden by a majority vote in both houses of Congress, for three months, Yucca Mountain was debated in congressional committee hearings, and on the floor of the House and Senate. Eventually, the House of Representatives voted to override the Nevada objection by approving the Yucca Mountain site 306 to 117. Later, the Senate concurred by a voice vote of 60 to 39. This approval, known as the Yucca Mountain Development Act (YMDA), was signed into law by President George W. Bush on July 23, 2002. The YMDA allowed the Department of Energy (DOE) to prepare and submit a license application to the Nuclear Regulatory Commission to move forward with the decision.

By the time the YMDA was enacted, the DOE had spent $7.1 billion on the issue; it would soon spend another $1.5 billion preparing the Yucca Mountain license application, including transportation and waste acceptance plans.[2]

Finally, in June 2008, the DOE submitted its staggering 8,600-page license application to the NRC. After a preliminary ninety-day screening period involving some forty NRC staff members, the NRC determined that the application contained sufficient information to formally move on to the next stage of technical and scientific review. According to federal legislation, the NRC was to complete the Yucca Mountain license application review within four years. According to the DOE, the earliest the repository could start accepting waste, given a smooth licensing process and consistent funding, was 2020.

But there was a problem. The four-year window could be slammed shut by the incoming antinuke president, Barack Obama.

Immediately upon taking office, Obama's first proposed budget cut off money to further ready Yucca Mountain for waste disposal. That move kissed the 2020 goal good-bye.

However, to silence political opposition and simultaneously appease the majority of Americans who are for nuclear power, in January 2010, Team Obama cleverly pulled the infamous bureaucratic trick of forming a blue ribbon commission to "provide recommendations for developing a safe, long-term solution to managing the Nation's used nuclear fuel and nuclear waste."[3]

The announcement made it appear as if the White House were being open-minded about reconsidering Yucca Mountain. However, in February 2011, proving the establishment of the committee was a sham, energy secretary Steven Chu unilaterally sent a letter to the blue ribbon team, declaring that the Yucca Mountain plan was *not* a "workable option."[4]

Obama and Chu's bluff was called three months later when the Government Accountability Office, Congress's independent investigative arm, released a statement saying, "[The] DOE's decision to terminate the Yucca Mountain repository program was made for policy reasons, not technical or safety reasons."[5]

To be sure, sixty-five thousand tons of anything sounds like a lot, but that amount of spent uranium does not require an expansive storage area. Extremely dense, a chunk of uranium the size of a gallon jug of milk weighs 150 pounds—so, thirteen gallon jugs of uranium would equal about a ton. Simple math would break down the uranium tonnage into roughly 867,000 milk jugs, all of which could fit neatly within the confines of the average high school basketball gym.

Just as he's done with our domestic oil supply, Obama has taken our nuclear power portfolio and heaved it backwards. Without prompt action, our nuclear sector will become so antiquated and our power grid so stretched that not only will our lifestyles be dumbed down, but our global economic standing will be further dented as the rest of the developed (and developing) world gains a significant advantage.

That's what Marxists have desired for decades.

Obama telegraphed his position on nuclear power long ago, and now we must be clear on the real issue: Obama, Chu, and their environmentalist cohorts do not want the American people to have access to inexpensive, plentiful power. As Paul Ehrlich once said, "Giving society cheap, abundant energy...would be the equivalent of giving an idiot child a machine gun."[6]

NUCLEAR POWER 101

There are currently 440 operating nuclear power reactors distributed around the globe in 47 different countries, accounting for 14 percent of the world's electrical production.[7] In Lithuania, 76 percent of the electrical grid is supplied by nuclear power; in France, it's 75 percent. In the United States, 104 nuclear power plants supply 20 percent of the electricity, with some states benefiting more than others. Worldwide there are 64 nuclear plants under construction; none of those are located in the U.S.[8]

One need not be a scientist to understand how electricity is generated in a nuclear plant, no matter how activists play on the general population's lack of atomic knowledge and relish in frightening the unenlightened with terms like "meltdown" and "radiation." They want such terminology to conjure up images of mushroom clouds and mutant life forms. However, once properly explained, it becomes apparent that the energy created in a nuclear power plant is actually really fascinating stuff.

Nuclear plants depend on the reaction that occurs during a process known as "nuclear fission," when one atom splits into two and subsequently releases energy (the splitting of an atom is why such power is often referred to as *atomic*). The energy released during nuclear fission produces a by-product known as *radiation*.

Nuclear fission is naturally occurring all around us every day. For example, uranium is a common element found virtually everywhere within the earth's crust, and certain types of uranium are constantly undergoing spontaneous fission at a very slow rate in a procedure know as *decaying*. Thus, as the element decays, it continually emits radiation. Because uranium is so plentiful, it's a natural choice for the key ingredient in the fission process that nuclear power plants require. While there are several varieties of uranium, uranium 235 (U-235) is among the most efficient for the production of nuclear power. More important, the U-235 atom is one of the few elements that can undergo artificially induced fission.

Let me take you back to middle school science for just a moment, where we learned that all matter is made up of atoms. Atoms are composed of *protons* and *neutrons*. A normal atom has a balanced number of protons and neutrons in its nucleus. For example, helium has 2 protons and 2 neutrons, and iron has 26 protons and 26 neutrons. If this balance were altered, the atom would become unstable and be referred to as an *isotope*. In a futile attempt to gain balance, isotopes are constantly in a state of decay, reducing themselves by half over and over again, and in the process, releasing radiation. This halving procedure is known as a *half-life*. The passage of about seven half-lives results in a reduction to less than 1 percent of a previous radioactivity level. Depending on the element, half-lives can vary from a fraction of a second to millions of years.

Now, back to uranium. About 99 percent of the uranium within the earth's crust is not U-235, but U-238. The two are both natural isotopes and chemically identical, but they differ in their physical properties, notably, their mass. The nucleus of the U-235 atom contains 92 protons and 143 neutrons, giving it an atomic mass of 235 units. The U-238 nucleus also has 92 protons but 146 neutrons—three more than U-235, and therefore, has a mass of 238 units. Through a technical process, the three extra neutrons from U-238 can be separated, thus creating a U-235 isotope. This procedure is known as "enrichment" and is how nuclear fuel is generally created.

Now, here's what is unique about U-235: it takes a long time for a U-235 atom to reach its half-life—hundreds of millions of years. However, this atom is exceptional in that fission can be artificially induced by way of a complex process conducted within the bowels of the reactor. So, instead of waiting millions of years for the U-235 atom to decay, nuclear scientists can artificially excite the element, forcing it to split immediately and causing its sudden decay and the release of energy in the form of heat.

The goal in a nuclear reactor is to set off a chain reaction whereby millions of U-235 atoms are simultaneously engaged in fission, which, in turn, produces a significant amount of heat from a relatively small amount of uranium. The heat created by the splitting atoms is then used to heat a water supply and make steam that spins a turbine that drives a generator, producing plentiful electricity.

The decay of a single U-235 atom releases approximately 200 MeV (million electron volts). While that figure may cause your eyes to glaze over, let me put it to you this way: if you were to apply this fission process to a pound of high-quality, enriched U-235 uranium (like the type you would find on a nuclear-powered submarine), the amount of energy would be equal to about a million gallons of gasoline. If you estimate you use ten gallons of gas each week, that means the equivalent of one pound of uranium could power thirty-eight cars fifty 50 years.

Now you understand why nuclear energy is remarkably efficient. It is a source of power sustained by a virtually unlimited fuel supply, capable of producing inexpensive energy 24/7 (the wholesale cost of which is about $17.60 per megawatt-hour or 1.76 cents per kilowatt-hour—significantly cheaper than coal and a quarter of the cost of natural gas–fired generation).[9] And nuclear electricity is baseload power that is *completely emission free*.

One more important note regarding enriched uranium: because the nuclear industry is federally regulated, the only company in the U.S. that holds a contract to enrich uranium is the United States Enrichment Corporation (USEC). USEC has been waiting on a $2 billion loan guarantee from the federal government to move forward with plans to turn an outdated enrichment facility in Piketon, Ohio, into a facility utilizing state-of-the-art technology. With Piketon's unemployment rate hovering near 15 percent in recent years, and despite Obama's big talk about creating jobs, the USEC's loan guarantee was postponed by the administration in 2009, and as of this writing, remains up for review. At stake are eight thousand jobs nationwide, half of them located in Ohio. Don't be surprised if the loan comes through at a strategic moment just prior to the 2012 election in an attempt to secure the electoral votes of Ohio, a key swing state.

INSIDE A NUCLEAR POWER PLANT

Uranium utilized at a power plant has been enriched some 2 to 3 percent (in comparison, a nuclear bomb requires 90 percent enrichment). The enriched uranium is typically formed into inch-long pellets, each approximately the same diameter as a dime. The pellets are arranged into "fuel rods," which are about twelve to seventeen feet in length. The rods are only mildly radioactive and can be safely handled with gloves. Thousands of fuel rods are arranged into bundles and clad in metal tubes made of a zirconium alloy. These bundles are submerged in water inside a durable pressurized containment vessel surrounding the reactor's "core." Once artificial fission is induced, the chain reaction of fission begins. The heat created through fission is used to warm the water, creating steam, which drives a turbine. The turbine's mechanical energy is then converted into electricity by means of a generator. There is no pollution or carbon dioxide emitted, only water vapor in the form of steam.

To keep the fuel rods from becoming too hot, a coolant is constantly circulating in the core. More important, if the chain reaction of fission begins to proceed at too great a rate, "control rods," composed, in part, of cadmium or boron (materials capable of absorbing neutrons), are inserted into the fuel rod bundles. By mechanically inserting the control rods, the fission process can be altered. If for any reason the reactor needs to be completely shut down (for maintenance, to change the fuel rods, or in the event of an accident), the control rods can be fully inserted into the uranium bundle, interrupting the fission process.

The fuel rods remain within the reactor for five years. When finally removed, they look exactly as they did on day one—like a bundle of metal pipes—except now they are quite radioactive. However, submerging them in a few feet of water or covering the rods with a sheet of lead will block all outgoing radiation; thus they can be stored safely in water or lead-lined casks. Within three years they will lose half their radioactivity.

It should be noted that the containment vessel housing the core is stalwart, functioning like the outer shell of an egg, protecting all the vital internal components necessary for fission to occur. For example, in a pressurized water reactor—the most common type used in the U.S.—the containment vessel is typically about forty feet tall and fourteen feet in diameter, with carbon steel walls some eight inches thick, surrounded by another four feet of reinforced concrete. The outer vessel alone weighs more than five hundred tons and is made up of seven massive steel forgings.

In addition to the containment vessel, there is the familiar dome-shaped tower that houses all the key components involved in the fission process, known as the *containment and missile shield*. This structure further seals the nuclear rods and associated components from the outside atmosphere. New construction standards are being employed to ensure that the containment shield is strong enough to withstand the impact of a fully loaded passenger airliner without rupture.

THREE MILE ISLAND

Everything in life involves a certain amount of risk; the key is mitigating that risk with safeguards as technology improves and becomes more affordable. When it comes to illumination, electrification, and energy, there is always going to be the risk of someone getting harmed, or worse. However, if you were to query the people of the world, without hesitation they would say the risks are worth the benefits because such technology greatly enhances life. Yet, in their zeal to save the planet, environmentalists would rather you use candlelight than a lightbulb. Such practice, however, would cause death and destruction, as each year in the U.S., candle usage causes more than 12,000 fires with 170 associated fatalities.[10]

But the green agenda is not about facts; it's about attacks, smear campaigns, and instilling fear in the hearts of otherwise decent people. Consider the unrelenting assault on the nuclear industry regarding a trio of iconic accidents: Three Mile Island; Chernobyl, Ukraine; and now Fukushima, Japan.

For decades, Three Mile Island (TMI) has been the prime illustration that the antinuke crowd loves to dust off to demonstrate that nuclear power is a threat to humanity. "There was a *meltdown* on U.S. soil," they whine.

Yes, there was a meltdown—actually a *partial* meltdown. That means TMI's nuclear rods overheated and fractionally melted before the control rods could be applied to halt the fission process and cool everything down.

The Three Mile Island generating facility is located on an island in the middle of the Susquehanna River, south of Harrisburg, Pennsylvania.

On March 28, 1979, TMI was the scene of the most serious nuclear power plant accident in U.S. history, even though there were no deaths or injuries to plant workers or members of the nearby community. The sequence of events—equipment malfunctions, design-related problems, and worker errors—led to the partial meltdown of Three Mile Island's second reactor core (TMI-2), with only minimal releases of radioactivity. The good news is that the industry learned a lot from the accident and as a result, greatly improved emergency response planning, reactor operator training, and radiation protection, which, in turn, enhanced overall safety worldwide.

TMI-2's problems began when the pumps feeding water to cool the core stopped running, allowing heat to abruptly accumulate in the containment vessel. As planned, the entire reactor automatically shut down within minutes. However, because of the heat buildup in the core, pressure began to rise rapidly. Next, a series of human blunders caused mechanical glitches that prevented adequate backup cooling measures to engage, resulting in the fuel rods severely overheating to the point at which the zirconium cladding surrounding the bundles began to melt. Before the control rods could be inserted and more coolant pumped in, about one-half of the uranium rods within the core melted.

Fortunately, the pressurized containment vessel surrounding the core worked perfectly. Its indelible carbon steel held tight with less than a fraction of an inch actually melting.

Despite the serious mishaps, disaster was averted because the in-depth safety systems functioned as planned and designed. Yes, there was a partial meltdown, but no, there was no breach of the vessel or giant containment building. Steam from the core was released through a ventilation stack, and radioactivity was expelled into the atmosphere. However, the quantity of radiation was minimal. In 1990, scientists from Columbia University's Mailman School of Public Health and the National Audubon Society in 1990 found no convincing evidence that the TMI accident caused a rise in cancer rates, confirming earlier findings by the Pennsylvania Department of Health.[11] Altogether, at least a dozen epidemiological studies were conducted and concluded that there were no discernible direct health effects to the population in the plant's vicinity.[12]

Nonetheless, to this day, groups such as Greenpeace use TMI as a scare tactic to prohibit the use of nuclear power. On the thirtieth anniversary of the TMI accident, the Greenpeace website dishonestly recounted the events from 1979, stating: "What happened was that a reactor went into melt down, and ten years later it was still too dangerous to approach. Some folks that lived in the area got large awards for possible damage. These awards were given in secret. One woman was given an award for having a Down syndrome baby."[13]

The truth is, greedy attorneys worked to score for "victims" of TMI for decades. Finally, in 2003, during the final chapter in a series of legal challenges, a federal appeals court dismissed the consolidated cases of two thousand plaintiffs seeking damages against TMI. The court ruled that the plaintiffs had failed to present evidence demonstrating that they had indeed received a radiation dose large enough to possibly cause health effects.[14]

Meanwhile, damages were paid—not in secret—to businesses and residents of the nearby community who incurred financial losses due to the precautionary evacuations that took place immediately after the accident. These payments demonstrated the effectiveness and integrity of the industry's liability insurance protection in accordance with federal law.

As for the fate of TMI-2: thousands of gallons of radioactive cooling agents were safely pooled in the reactor building's basement and auxiliary storage tanks. Workers entered the reactor building soon after the accident, but the basement was not to be entered until a plan (and the technology to execute that plan) could be successfully developed. That took several years. By 1990, though, the basement had been cleared of all radioactive material.

Though TMI-2 was never reopened, the Nuclear Regulatory Commission has deemed it "decontaminated to the extent the plant is in a safe, inherently stable condition suitable for long-term management."[15]

CHERNOBYL AND FUKUSHIMA

Likening TMI to the disaster at Chernobyl is like comparing apples to artichokes.

On April 26, 1986, an accident occurred at Unit 4 of a nuclear power station at Chernobyl, Ukraine, in the former USSR. A worker at the plant violated protocols, allowing a reactor core to suddenly become far too hot, creating a powerful explosion that destroyed the entire reactor and disbursed massive amounts of radioactive material into the atmosphere. It should be noted that the containment vessel surrounding the core was pathetically weak by U.S. standards and there was *no* containment building, let alone a missile shield.

The Chernobyl tragedy caused many severe radiation effects almost immediately. Among the approximately 600 workers present on the site at the time of the accident, 2 died within hours of the reactor explosion, and 134 suffered from acute radiation sickness. Of these, 28 workers died within four months of the event.[16] The health of the Chernobyl residents has been monitored since 1986, and to date there is no strong evidence for radiation-induced increases of leukemia or solid cancers. However, about 4,000 cases of thyroid cancer were detected among the youth, with 99 percent of these children successfully treated.[17]

Thankfully, shoddy nuclear plants like Chernobyl died along with the Soviet Union.

Altogether different was the perfect storm of trouble that hit the Fukushima, Japan, nuclear plant on March 11, 2011. While the nuclear reactors at that facility were able to withstand one of the strongest earthquakes ever recorded—9.0—the facility was not able to stave off an unimaginable forty-nine-foot tsunami, which easily rolled over the facility's nineteen-foot seawall.

At the time of the earthquake, three of the plant's six reactors were turned off for planned maintenance. When the earthquake struck, the three remaining reactors automatically shut down on cue, immediately shutting off electricity to the plant. Per design, backup diesel generators instantly kicked in, allowing the plant's cooling systems to function properly and keeping the rods from overheating.

This should have been the end of the story. The Fukushima reactors were no longer in fission mode, no electricity was being produced, and the fuel rods were being cooled.

However, forty minutes after the quake, all hell broke loose. The enormous force of the tsunami smashed into the facility and swamped the generators, killing the primary auxiliary power to the plant. The secondary alternative electrical source—eight hours of battery power—also failed. At this point, there was no contingency plan. The uranium fuel rods, enclosed in the zirconium tubes, became scorching hot. Under such conditions zirconium reacts with water to form zirconium oxide and hydrogen. The hydrogen was being vented into the containment buildings and quickly created immense pressure, producing dramatic explosions and compromising the containment structures. Fortunately the containment vessels surrounding the reactor cores in each of the three operating plants were *not* damaged. According to the Nuclear Energy Institute, "Although some fuel melting may have occurred, there is no evidence of complete core damage at any reactor."[18]

Eventually additional power was brought into the facility, but by then the fuel rods within the reactors had been terribly compromised and a partial meltdown was realized with radiation having been emitted into the atmosphere.

Despite many stories to the contrary, no one died from radiation exposure in Japan. Two workers at the plant were killed, but from conditions associated with the earthquake.

Three months after the tsunami, congressional hearings were held in Washington, D.C. John Boice, one of the world's top radiation epidemiologists (who happened to be in Hiroshima, Japan, on the day of the earthquake, reviewing a study of World War II atomic bomb survivors), presented testimony. Boice has spent his career analyzing human exposure to natural and man-made radiation and determined that the health consequences for the public due to the Fukushima nuclear plant accident were minimal. Boice further explained that a small number of workers, including two that stepped into contaminated water, received doses of radiation that could *slightly* increase their lifetime risk of developing cancer. However, regarding the general public, Boice concluded, "Thus, while Fukushima is clearly a major reactor accident, the potential health consequences associated with radiation exposures in terms of loss of life and future cancer risk are small."[19]

In the days after the Fukushima explosions, I utilized a Geiger counter around San Francisco in an attempt to detect fallout from Fukushima. While I was unable to notice such particles in the atmosphere, there were reports of slightly elevated radiation levels in the Pacific Northwest. Regarding this, Boice stated this radiation was thousands of times below government limits. "The tiny amounts of detected radioactive materials from Fukushima pose no threat to human health. They represent, at most, only a tiny fraction of what we receive each day from natural sources, such as the sun, the food we eat, the air we breathe and the houses we live in."[20]

HOW MUCH RADIATION?

Recall this story?

During the experimental detonation of a gamma bomb, scientist Bruce Banner rushes to save a teenager who has driven onto the testing field. Though able to heroically push the unwary teen into a ditch, Banner catches the direct force of the atomic blast, absorbing massive amounts of radiation. He later awakens in an infirmary, seeming relatively unscathed. However, that night Banner transforms into a lumbering, muscular, green form that breaks through the wall and escapes. A soldier in the ensuing search party dubs the mysterious creature "the Hulk."

We've grown up with some creative stories regarding radiation, and the antinuke crowd tries to take advantage of our apprehensions at every opportunity, throwing the R-word—*radiation*—around as if it's a flimsy bag filled with deadly virus. The truth is, radiation is a natural process that is occurring at all times all around us. As I mentioned, I'm one of those geeky guys who owns a Geiger counter. Whenever I turn it on, folks are amazed at all the radioactivity it detects from materials that can be broken down into three basic categories: *cosmic*, *terrestrial*, and *technological*.

Cosmic radiation emanates from outer space and the sun; *terrestrial radiation* is found in soil and rocks, and thus can be identified in common building materials; and *technologically enhanced radiation* is generated from X-rays, cancer-fighting agents, and nuclear plants.

Radiation is measured in units called *millirems* (mrems). The average person experiences a dose of about 620 mrems per year. International Standards consider exposure to as much as 5,000 mrems a year safe for those who work with and around radioactive material.

Here is an example of how much radiation you are exposed to annually:[21]

COSMIC RADIATION

- If you live at sea level, you're exposed to 26 mrems; 2000–3000 feet elevation, 35 mrems; 5000–6000 feet elevation, 52 mrems.

- Jet plane travel. For each hour you spend in the air add 0.5 mrem per hour traveled.

TERRESTRIAL RADIATION

- If you reside near the Gulf or Atlantic coast, 16 mrems; Colorado Plateau, 63 mrems; anywhere else in the continental U.S., 30 mrems.

- If you live in a stone, brick, or concrete building, 7mrems.

- Coal plants. If you live within 50 miles of a coal-fired electrical generating plant add .03.

- False teeth. Add .07mrems for each porcelain crown or false tooth.

- Glow-in-the-dark dial on your wristwatch. Add .0006

- If you smoke a half-pack of cigarettes each day of the year, add 18 mrems.[22]

TECHNOLOGICAL

- X-rays. Chest 10 mrems, dental 0.5; abdomen 700, spine 600; barium enema 800.

- CT Scan. Head 200 mrems; heart 2000, spine/whole body 1000.

- Each time you walk past a luggage X-ray machine at the airport add .002 mrem. When you go through a TSA full-body scanner (back-scatter x-ray machine) you receive .009 mrem.[23]

- Nuclear power plants. If you live within 50 miles of a nuclear plant add .01.

As you can see, the radiation to which a community is exposed from a nuclear power plant is negligible, especially compared to other common sources of radioactivity. According to the Nuclear Energy Institute, the average U.S. nuclear power plant worker receives 120 mrems annually. Based on the track record of the nuclear power industry, a competently run, well-maintained nuclear power generating facility is a safe environment that provides the public with a brilliant source of energy. Its problem is that the eco-socialists are winning the PR battle—especially in America.

REDUCE, REUSE, RECYCLE

Over the last fifteen years, environmentalists have successfully stamped a slogan into the minds of America's students: "Reduce. Reuse. Recycle." These verbs are also known to eco-educators as, the "Three Rs of the waste hierarchy," or as the Obama administration demands, "Three great ways YOU can eliminate waste and protect your environment."[24]

In fact, when it comes to the waste hierarchy, Obama's director of the National Institute of Environmental Health Sciences (NIEHS), Linda Birnbaum, has taken the instilling rhetoric to a new level. The "Kids' Pages" on the NIEHS website reads like a Marxism primer: "First and foremost, buy and use less! If all the other people on the Earth used as much 'stuff' as we do in the United States, there would need to be three to five times more space just to hold and sustain everybody... WOW!"[25]

Reduce, reuse, and recycle. Just do it and we'll have "more space," they claim. It's too bad Team Obama refuses to employ the Three Rs when it comes to nuclear waste. Right now we have the technology to *reduce* the amount of nuclear waste through the *reuse* of the uranium in the fuel rods, which could happen if the government would allow us to *recycle* those used nuclear fuel rods.

The tragedy is that the very ideologues who hold the Three Rs in higher esteem than the Ten Commandments are preventing U.S. nuclear facilities from being able to employ those same rules. If the government is so hot-to-trot on the so-called Three Rs, they need to walk the talk and apply their sophomoric mantra to the radioactive waste generated by the nuclear power industry!

It's maddening to think that after a single use in a nuclear reactor, approximately 97 percent of the recoverable uranium and plutonium (plutonium is a by-product of the fission process in the reactor's core) from spent nuclear fuel is classified as non-recoverable waste.[26] That means that nearly all of the fuel in the spent bundle of rods still contains energy value, and yet the government won't allow the fuel to be repurposed and used again. Instead, after one service, the nuclear fuel rods are permanently stored in a repository. In the nuclear industry, this is known as an "open" fuel cycle, as opposed to the recycling and reuse of nuclear rods, which is a "closed" fuel cycle. Through recycling and reuse, precious uranium could be sorted out to become new fuel and the volume and toxicity of the waste to be stored greatly reduced. If closed fuel cycles had been employed from the beginning, only a fraction of the current sixty-five thousand tons of waste would need to be permanently housed in a repository.

HALTING NUCLEAR FUEL RECYCLING

Ironically, U.S.-based companies developed the technology to recycle spent fuel rods decades ago and were carrying out operations in three facilities, but then in 1976, their brilliant technology was politicized and ground to a halt.

On October 28, 1976, less than a week before the 1976 presidential election, antinuke activists launched a flash campaign to shut down the reprocessing industry. They claimed to be concerned about terrorists stealing plutonium (the preferred fissile material for building a nuclear bomb) from nuclear fuel reprocessing plants and then using the material to proliferate a weapon of mass destruction.

In a knee-jerk political response to quickly quench those far-fetched machinations, President Gerald Ford announced a temporary halt to nuclear fuel reprocessing, hoping he could score points with undecided voters.

The ensuing vote was a close one, but liberal Jimmy Carter was elected president, and four months after taking office, he extended the temporary reprocessing moratorium into a long-term policy, thus deferring indefinitely the commercial reprocessing and recycling of spent rods from commercial U.S. nuclear power plants (the military continued to recycle their spent fuel).

In October 1981, Carter's predecessor, Ronald Reagan, announced he was "lifting the indefinite ban which previous administrations placed on commercial reprocessing activities in the United States."[27] However, there was a complication—by 1981, there were no facilities in operation to recycle spent fuel rods; they had all closed down. Additionally, the nuclear industry was not interested in assuming the risk to spend billions on such an operation, which might be shuttered by a future antinuke president.

Sure enough, by 1993, President Bill Clinton reinstated the Carter policy that to this day has denied reprocessing and discouraged research.

Other countries have not taken such a senseless approach to nuclear power. France, for example, has safely recycled nuclear fuel for decades. Upon removal from French reactors, the used rods are packed in containers and safely shipped via train and highway to a facility in La Hague. There, viable uranium and plutonium are removed and separated from the nonrenewable materials and processed into new rods. The French have recycled nuclear fuel for thirty years without incident and have now safely processed more than twenty-three thousand tons of used fuel—enough to power France for fourteen years. France is even using their plant in La Hague to recycle spent nuclear material from other countries.

The UK, Japan, India, and Russia all engage in some level of reprocessing. China is moving in this direction too.

The real problem with nuclear waste in the U.S. is that our government has allowed environmentalists to keep us from employing our own made-in-the-USA technology to recycle nuclear fuel. Their motivation is perverted: they look forward to the day when we have so much nuclear "stuff" that we won't know what to do with it. Simultaneously that stuff is used to scare the public into believing there is no such thing as a beneficial nuclear power plant.

AND THE TAXPAYERS GET THE BILL

Meanwhile, American taxpayers are getting charged for a national nuclear storage plan that has never manifested.

Since 1982, in compliance with federal law, consumers using the electricity produced by America's nuclear plants have paid a fee of one cent per kilowatt-hour of power generated to bankroll the Nuclear Waste Fund waste depository. As noted, in 1987 it was determined that such a depository would be located in Nevada. Thus far, a whopping $30 billion has been set aside for this endeavor. However, since Yucca Mountain is now on ice, the fee-payers, us, should get a refund. And I'm not the only one who feels this way—so do the utility companies that run the nuclear plants.

Utilities are seeking to recover billions of dollars they have spent over the last two decades on storing spent nuclear fuel and radioactive waste while waiting for the government to construct a permanent storage facility. The utilities were obligated by Congress under the Nuclear Waste Management Act of 1982 to temporarily store the waste while a permanent storage facility they would help fund was built in Nevada. Currently spent nuclear fuel is held at 122 temporary sites in thirty-nine states. The costs to the utility companies, and therefore the consumer, are immense. To date, some seventy lawsuits have been filed against the DOE regarding the government's inability to go forward with the promised permanent waste facility, and some of the settlements have been appropriately significant. In March 2010, the U.S. Court of Federal Claims awarded Energy Northwest nearly $56.9 million in damages in past costs for spent nuclear fuel storage; in 2004, the DOE settled a lawsuit by the Exelon Corporation, the nation's largest nuclear power plant operator, for $80 million. Assuming Yucca Mountain remains closed, Exelon will get $600 million through 2015.[28] All told, the potential liability for all of the pending lawsuits could be more than $60 billion.[29]

100 NEW NUKES

In 2005, the chairman of the Federal Nuclear Regulatory Commission, Dr. Nils Diaz, recommended building 100 American nuclear power plants.[30] Globally, since 2000, 22 new nuclear plants have been put on line, and some 25 more are under construction now. Amazingly, the last unit to be turned on in the U.S. was the Watts Bar Unit 1 reactor in East Tennessee in June 1996. The last successful order for a U.S. commercial nuclear power plant was in 1973—nearly forty years ago.

In 2009, President Obama designated Gregory Jaczko as the new chairman of the Nuclear Regulatory Commission. The anti-nuke crowd regards Jaczko, a former advisor to green-friendly senators Ed Markey and

Harry Reid, as a hero. In October 2011, as the commission was in the final phase of deciding to license the first new nuclear reactors to be built in the U.S. in decades, Jaczko's fellow commissioners sent a letter to the White House expressing "grave concern" about Jaczko's obstructionist actions. Finally, in February 2012, when the votes were cast to approve two reactors at Georgia's Vogtle nuclear plant, Obama's man Jaczko was the lone "no" vote. Nonetheless, the 4 to 1 decision by the NRC will allow the U.S. to take a baby step in the right direction toward energy security.

The federal government needs to stop catering to the concerns of the radical, uncompromising environmental lobby and allow the nuclear industry to construct new facilities in the U.S. Doing so will mean a stable power grid, reduced energy costs, significant job creation, and continued competitiveness with the rest of the world.

In 2009, Republican senator Lamar Alexander of Tennessee asked, "So why not build 100 new nuclear power plants during the next 20 years? American utilities built 100 reactors between 1970 and 1990 with their own (ratepayers') money. Why can't we do it again? Other countries are already forging ahead of us."[31]

The senator makes an excellent point. There is no coherent reason we cannot replicate what we've already accomplished. We are a nation of doers. In 1961, President John F. Kennedy called for a moon landing by the end of the decade. In eight years we went from launching a rocket carrying a chimpanzee to watching men walk on the moon. There's no reason we can't build 100 nuclear power plants in the next twenty years, and then keep on constructing them.

Census data indicates that the U.S. population will rise by some 50 million people come 2032; that's a 16 percent increase. By 2050 the population will expand by 100 million, or roughly 33 percent from today. Given the successful push by environmentalists to tear down hydroelectric dams and prevent new coal-fired electrical generating plants from being constructed, simply to avoid rolling blackouts and incredibly expensive energy from becoming the norm, we need to begin constructing a fleet of new nuclear power plants *now*. As we've discussed, wind and solar are not going to provide baseload electricity—they'll only help reduce peak energy demands. Thus, just to maintain parity within our nation's energy portfolio, a goal of having 200 nuclear power plants by 2032 will likely provide America with about 20 percent of its energy output—roughly the same amount that nuclear energy provides today. Forecasting to the year 2052, those same 200 plants would be producing only 14 percent of U.S. electricity. In other words, we must plan ahead if we want to meet the needs of our citizens and provide the basic energy infrastructure to support our industrial base.

We need more nuclear power plants.

In 1979, prior to the TMI accident, there were 150 new reactors on the drawing board, with most of the sites already chosen. Because of TMI, only 50 of these were ever opened. Building another 100 reactors by 2032 would simply mean going back and finishing the job.

Many lessons have been learned in the meantime. Almost every reactor built in the 1970s and '80s was unique. This was likely one of the flaws that led to TMI. Because no two reactors were the same, the industry was void of a common pool of knowledge and did not communicate well regarding safety. Today the manufacturers take a completely different approach. Each company has concentrated on a standard model, allowing the NRC to issue "design certifications" for those models so each utility applicant does not have to start again from square one.

One hundred new nuclear facilities will cost a lot of money but should not be prohibitively expensive. Current high-end estimates are that a new reactor costs $5–$7 billion. A price tag in this range would put the price of 100 new reactors around $700 billion, less than the cost of the 2009 Stimulus. However, in this case, all the money could originate with the private capital, because with regulatory uncertainties removed, nuclear power becomes an attractive investment. While coal prices are steadily rising, natural gas prices always in flux, and wind and solar only feasible with government subsidies, nuclear fuel prices are locked up in long-term contracts; thus, reactors currently are able to garner a profit of close to $2 million a day selling electricity. Once up-front construction debts are paid off, these profits help maintain low rates and provide made-in-America jobs.

And the cost of this electricity will be cheap. According to the Nuclear Energy Institute, the production cost per kilowatt hour for electricity derived by natural gas is 11 cents; coal, 5 cents; nuclear, 4 cents; and hydro, 3 cents.[32] A major nuclear construction program would be a no-brainer move for consumers.[33]

In terms of government-aided funding, how about the plan created in 2010 by Senator Pete Domenici (R-NM), who proposed the creation of a Clean Energy Bank, a government corporation equipped with $100 billion in loan guarantee authority? This new institution would be modeled on the United States Export-Import Bank and the Overseas Private Investment Corporation. I've been quite critical of these in-place entities because Team Obama has used funds from the Export-Import bank to fund $2 billion worth of oil exploration for Brazil's state-owned oil company, Petrobras, and another $1 billion to help Brazil build stadiums for the 2014 World Cup and the 2016 Olympic games. If our federal government is able to loan money for such foreign projects, how about ensuring that capital flows to critical infrastructure deployment in the U.S. electric sector?

REAL JOBS

One hundred new reactors would be a boon to American industry and would create thousands of excellent, permanent jobs.

According to Admiral Frank Bowman of the Nuclear Energy Institute, peak employment per reactor during construction could be as many as 3,000 to 4,000 full-time jobs.[34] Assuming it takes five years to build a nuclear plant, if 20 such facilities were built every five years, that's 60,000 to 80,000 jobs per five-year construction cycle. Once the plants are online, there will be at least 700 permanent, high-paying jobs with great benefits—real "green" jobs that cannot be shipped offshore. Each plant would then naturally spur many more employment opportunities in the surrounding community, providing goods and services necessary to support the primary workforce.

There's another employment factor here that cannot be overlooked. Creating a reactor containment vessel is a job that demands superior iron-forging capabilities. There are nearly two hundred forgings of various sizes, weights, and specifications required for today's newest generation of nuclear power plants. However, while there are dozens of manufacturers capable of providing small-to-medium-sized forgings, no company in the U.S.—once the world's steel leader—is capable of delivering the ultra-heavy forges required to fashion a nuclear containment vessel. Four of the most complex parts of a nuclear power plant—the containment vessel, reactor vessel components, turbine rotors, and steam generators—are made from more than four thousand tons of steel forgings.

"The real crux of the issue in the United States comes down to whether we want to produce these large parts," says Charlie Hageman, executive vice president of the Forging Industry Association. "It's a supply chain issue. And what we see is that the government policy doesn't appear to be clear and the companies that would make the large parts are reluctant to invest in adding to their capabilities."[35]

Presently, only one company in the world, Japan Steel Works, is in the business of forging and exporting reactor vessels, and there is a four-year waiting list for their services. Russia and China (China is preparing to build 60 reactors) have now completed new steel forges so they do not have to wait in Japan's line.

If we got involved in this market, we could own it. And we're talking about thousands of excellent, permanent American jobs.

If President Obama would pull the trigger to expand our nuclear portfolio and reinstate our ability to recycle our nuclear fuel, all told, tens of thousands of Americans could find honorable careers. On top of this, each year, the average nuclear plant generates approximately $430 million in sales of goods and services in the local community and nearly $40 million in total labor income. That's thousands more jobs.

And then there's total state and local tax revenue of almost $20 million from every plant to benefit schools, roads, and other state and local infrastructure, which equate to more jobs. Add to that annual tax payments of roughly $75 million per plant to the federal government, and you have billions to use to pay down our exploding national debt.[36]

America clearly needs a sound energy plan. We need to stop the rabid environmentalists who are purposefully instituting policies and laws aimed at decimating our power portfolio, including yet another form of emission-free electricity: hydroelectric dams.

ELEVEN

DAM IT

YOU CAN'T LIVE WITHOUT WATER. Three or four days without it and you're dead.

You can't grow food without water either, and rainfall alone isn't reliable enough to sustain major food production. That's why irrigation, a form of human intervention, is necessary to prevent food shortages and starvation—even in the United States.

If we are to meet the fundamental needs for survival as a nation, especially as our population increases by 100 million residents by the year 2050, then we need significantly more water, and we need to begin expanding our water supply *immediately*. We need more water for drinking, farming, ranching, and manufacturing.

We also need more dams, which create inexpensive, emission-free electricity.

The problem is that environmentalists are determined to maintain a limited water supply as an effective tool to restrict development, devalue private property, and therefore, better control our lives. They despise dams and find reservoirs repugnant; in fact, they are using the same tactics they've used against the oil and nuclear industries: junk science, ad hominem arguments, and throwing hissy fits to prevent new dams and reservoirs from being constructed. Worse yet, they are on a tear to deconstruct as many dams and reservoirs as possible in the name of "restoring the environment."

Thus far, the radicals are winning this water war. According to federal estimates, at least thirty-six states are on the verge of experiencing critical water shortages.[1] Each time a city, county, state, or member of Congress proposes a new water storage plan, the environmentalists leap into battle mode and almost always win.

This chapter will bring us a step closer to revealing the eco-tyrannists' plan for America, a plan in which humans are herded into urban hubs and vast expanses of land are permanently placed off-limits to development and recreation. Their design is to return North America to what they believe it resembled when native Americans trod lightly upon the face of Gaia.

Abundant water storage does not fit into that plan.

Humans have been building dams and reservoirs since the earliest days of recorded history—even native Americans were known to have constructed them. One of the earliest recorded dams was built on the Nile River about 2700 BC to provide water for Memphis, the ancient capital of Egypt. The Grand Anicut dam in South India, constructed of massive unhewn stones stretching more than three hundred meters across the Cauvery River, was built in AD 150 and to this day aids in the irrigation of a million acres of agricultural land. Humans have also been harnessing water to perform work for thousands of years. The Greeks used water-powered paddle wheels to move millstones that ground wheat into flour more than two thousand years ago. Similar mechanisms were designed to move the blades necessary to saw lumber, and power the mills to fashion textiles.

Throughout history, mankind has utilized its intellect and determination to use nature's raw materials to wonderfully enhance life. Dams and reservoirs are a perfect example. They store precious rainwater and snow runoff during wet seasons for use in the dry seasons; they reduce the risk of disastrous flooding; they assist navigation on major rivers by leveling them during high flow and low flow periods; they create places for recreation, tourism, and business; and they produce power and electricity by tapping the energy of falling water.

DAMS FOR AMERICA

The widespread construction of today's large dams did not begin until the early 1900s, when improvements in engineering, technology, and construction skills made it possible to build them safely. Initially the U.S. proved the leader in large-dam construction, with the government building about fifty such dams between 1900 and 1930. The first of these massive projects was the Hoover Dam, by the U.S. Bureau of Reclamation, on the Colorado River southeast of Las Vegas, Nevada, in the 1930s. By 1945, the five largest dams in the world had been completed, all in the western United States. Between 1930 and 1980, thousands more were constructed of varying sizes. Today, dammed reservoirs store 60 percent of the nation's entire average annual river flow.[2]

And then there is the electricity generated by dams. The evolution of the modern hydropower turbine began in the mid-1700s when a French hydraulic and military engineer, Bernard Forest de Bélidor, wrote *Architecture Hydraulique*. During the 1700s and 1800s, water turbine development continued to advance, and by 1880 the U.S. saw her first hydroelectric dam constructed in Grand Rapids, Michigan. More electricity-producing dams soon followed. In 1881, a water-spun turbine at a flour mill provided street lighting in Niagara Falls, New York. A year later, the Appleton Paper & Pulp Company harnessed the Fox River in Appleton, Wisconsin, to turn waterpower into electricity to power their large paper mills.

In 1896, Nikola Tesla and George Westinghouse became the first to design a hydroelectric power plant to distribute power on a wide scale. The Niagara Falls power plant in New York sent electricity to Buffalo, a distance of about twenty-six miles. By the early 1900s, more than 40 percent of U.S. electricity was produced by hydroelectric power; in the 1940s, hydroelectricity accounted for about a third of the country's electricity supply. Today hydropower accounts for a mere 7 percent of electric energy produced in the United States; worldwide the figure is closer to 24 percent.[3]

It's easier to build a hydro plant where there is a natural waterfall. That's why it was easy for Tesla and Westinghouse to select Niagara Falls. If a waterfall is unavailable, then constructing a dam, which essentially creates an artificial waterfall, is the next option. Dams are built on rivers where the terrain will allow the creation of an artificial lake or reservoir above the dam. Today there are about 80,000 dams in the United States, but only 2,400 are used to generate power.[4] Most dams are built for flood control and irrigation, not to produce electricity.

A dam serves two purposes at a hydro plant. First, the dam increases the height of the water. Second, it controls the flow of water. To generate electricity, a dam opens its gates to allow water from the reservoir above to flow down through large tubes called *penstocks*. At the bottom of the penstocks, the fast-moving, falling water spins the blades of turbines. The turbines are connected to generators that produce electricity. The electricity is then transported via huge transmission lines to a local utility company.

Currently there are about 80 million kilowatts of hydro-generating capacity in the United States.[5] That's equivalent to the generating capacity of 80 large nuclear power plants.[6] The biggest hydro facility in the U.S. is located at the Grand Coulee Dam on the Columbia River in northern Washington State.

Dams are the cheapest way to generate electricity. No other energy source, renewable or nonrenewable, can match them, because there is no cost for fuel. Hydroelectric plants also have long economic lives, with plants operating from fifty to a hundred years. Operating labor costs are also low, as the plants are automated and have few personnel on site during normal operation. It costs about one cent per kilowatt-hour (kWh) to produce electricity at a typical hydroelectric plant[7] (as opposed to 2.14 cents for nuclear, 3.06 cents for coal, and 4.86 for natural gas, as detailed in chapter 7).

Because the most economical sites for hydroelectric dams have already been developed, the construction of additional grand-scale projects are unlikely. However, existing plants could certainly be enlarged to provide further generating capacity, and many flood-control dams not equipped for electricity production could be retrofitted with generating equipment. A 2006 study by the U.S. Department of Energy's Idaho National Laboratory found 130,000 stretches of stream around the country suitable for small hydro projects, defining small as those between 10 kilowatts and 30 megawatts. Theoretically, if they were all developed, they could provide about 100,000 megawatts of power.[8]

However, despite the fact that hydroelectric power is 100 percent emission-free, the environmentalists—particularly in America—desire to see all of these plants shuttered and the associated dams removed in an effort to restore the land. If they have their way, economic development in America will cease and the population will be decreased.

ZERO POPULATION GROWTH

Those pushing to dismantle our dams and reservoirs believe the earth's population has surpassed the point of no return. Led by a host of socialist academics nurtured by the likes of Paul Ehrlich, they believe the earth, currently inhabited by nearly 7 billion people, can only sustain about 2 billion. According to Ehrlich:

> The current population...is being maintained only through the exhaustion and dispersion of a one-time inheritance of natural capital including topsoil, groundwater, and biodiversity. The rapid depletion of these essential resources, coupled with a worldwide degradation of land and atmospheric quality indicate that *the human enterprise has not only exceeded its current social carrying capacity, but it is actually reducing future potential* biophysical carrying capacities by depleting essential natural capital stocks"[9]

Ehrlich and his fellow travelers believe lifestyles must be altered to sustain the population. Their preferred method is social engineering via government rules, regulations, and laws. They also believe it's paramount that humans are regarded on a plane equal to that of a mere animal—not as the superior species. Such thinking makes the rationalization to decrease the human population more excusable. According to Ehrlich, if the proper policies are implemented,

> people will alter their lifestyles and thereby reduce their impact. Although we strongly encourage such changes in lifestyle, *we believe the development of policies to bring the population to (or below) social carrying capacity requires defining human beings as the animals* now in existence... The more important window may thus be a political one for laying the institutional foundations for desired change.[10]

What Ehrlich is proposing is a dangerous, time-tested ploy often instituted prior to government-sponsored human persecution and, in extreme cases, elimination. Ehrlich's former writing partner, the aforementioned John Holdren—President Obama's science czar—has concurred with Ehrlich, stating that humans are "just one species" in a world in which they "are not treated specially."[11]

Demoting the human species from superior to animal-like, or even just run-of-the-mill, is a horrific step in the wrong direction. In Hitler's Germany, Jews were considered imperfect and nearly subhuman. With the help of a government-funded PR campaign emphasizing the status of Jews, it became excusable for too many in Germany to accept the mandatory roundup of the Jewish population and allow them to be exterminated like vermin. Likewise, today in Africa, the black Christian and animist tribes of southern Sudan are being slaughtered by the Arab Muslim tribes of the north because the accepted teaching among the Muslims is that the southern Sudanese are a subordinate class and worthy of being hunted like animals. This is how mobs rule: select a target, demonize it, denigrate it, and then eliminate it. How can five billion people be subtracted from the global population unless one is willing to accept massive waves of death? Simply slowing down the birthrate is not enough. Natural disasters, such as earthquakes, tsunamis, hurricanes, and monsoons are a start. Disease is an effective tool, but starvation is certainly quicker; and war, though expensive and messy, may be the most effective precept of all.

These are maddening thoughts, but such bizarre conversations occur regularly among radical environmentalists and academics. Take, for example, Professor Joseph Tainter of Utah State University. Tainter is a very popular speaker and author of numerous books, articles, and papers on the subject of collapsed societies. Tainter says there are two ways to reduce the human population—catastrophe, the likes of which I've described, or severe economic hardship:

> [Either through a catastrophic] "crash" that many fear—a genuine collapse over a period of one or two generations, with much violence, starvation, and loss of population. The alternative is the "soft landing" that many people hope for—a voluntary change to solar energy and green fuels, energy-conserving technologies, and less overall consumption. This is a utopian alternative that...will come about only if severe, prolonged hardship in industrial nations makes it attractive, and if

economic growth and consumerism can be removed from the realm of ideology.[12]

Notice that the so-called soft-landing scenario is regarded as utopian, but it does make you realize how our current national and global economic problems are seen as opportunities to push the Green Agenda.

In his book *The Collapse of Complex Societies*, Tainter argues that a collapse is an economizing process carried out by rational individuals for whom this outcome is objectively preferable. He also warns that historically, the collapse comes quickly. "The process of collapse," he states, "is a matter of rapid, substantial decline in an established level of complexity. A society that has collapsed is suddenly smaller, less differentiated and heterogeneous, and characterized by fewer specialized parts."[13]

The crash theory has actually been in vogue with the believers of a green world for years. In 1988, science fiction writer Isaac Asimov told Bill Moyers of PBS Television that because of the growing population problem, "human dignity cannot survive. Convenience and decency cannot survive. As you put more and more people onto the world, the value of life not only declines; it disappears. It doesn't matter if someone dies, the more people there are, the less one person matters."[14]

Al Gore has advocated a reduced population for years as well, even stating the need for "sacrifice, struggle, and a wrenching transformation of society."[15]

During a June 20, 2011, keynote address at the Games for Change Festival in New York City, Gore said society must "stabilize the population, and one of the principal ways of doing that is to empower and educate girls and women."

According to Gore, there's a name for such education—"fertility management": "You have to have ubiquitous availability of fertility management so women can choose how many children to have, the spacing of the children." Gore says the goal is to make "parents feel comfortable having small families and most important—you have to educate girls and empower women."

One would guess that included in Gore's portfolio of fertility management is on-demand abortion services paid for by the government.

Like the eco-religious zealots described in chapter 6, the soothsayers of doom pushing a reduction of the planet's residents have the ability to whip people into a state of panic, forcing them to conclude that procedures must be instituted to engineer the world into a greatly reduced state. In the meantime, these same so-called guardians of ecology prefer pathetic mosquito nets over effective DDT in an attempt to appear concerned about death by malaria in Africa; to hand out condoms like Mardi Gras beads instead of teaching abstinence and fidelity to supposedly fight AIDS; and to advocate the use of waterless latrines and trickling wells as opposed to infrastructure that includes sewers, plumbing, reservoirs, and water treatment to prevent dysentery and other deadly diseases. They don't want progress; they want misery. They don't wish for quality of life; they wish for less life—less human life. Withholding water from the masses is the ultimate form of control; and certainly in the U.S. it will force lifestyle changes and convince some to have fewer children, or perhaps none at all.

THE MAN WHO WATERED THE WEST

While the environmentalists lack empathy for the needs and desires of humanity, there have been others who have addressed these requisites in significant ways; perhaps the most prominent of those is Floyd Dominy. If you live in the metro areas that include Phoenix, Las Vegas, or Los Angeles, you need to thank Floyd Dominy each time you turn on the tap. Likewise, if you eat any of the delicious vegetables, fruits, and nuts produced in California (believe it or not, agriculture is the Golden State's leading industry), then you also need to give a cheer to Mr. Dominy.

Frank Dominy joined the U.S. Bureau of Reclamation in 1946 and became its commissioner in 1959, a position he held for ten years. He was the man who implemented the plan to provide the Southwest with the water necessary to turn a desert into a prosperous land of opportunity.

Eco-socialists *hate* Floyd Dominy to such a degree that books and documentary films have been produced with the sole intention of smearing his character. Without Dominy's tenacity, the Southwest would be an impoverished stretch of sand habitable only by snakes, lizards, and indigenous peoples.

Dominy was born on a 160-acre sustenance farm near Hastings, Nebraska. In 1958, while testifying before Congress, Dominy explained the challenges of growing up on such a plot of land. "On that farm six of us children were born and six of us reached maturity on the subsistence of that 160-acre homestead. We had outside plumbing. We did not have deep freezers, automobiles, [or] school buses coming by the door. We walked to school in the mud. We had...one decent set of clothes to wear to town on Saturday; otherwise we wore overalls."[16]

In the 1930s, Hastings was located in one of the nation's most drought-stricken regions. "So I knew the value of water," Dominy said in a 2000 radio interview at the age of ninety. "I knew the problems that farmers had in making a living in the arid West. I had all that in my background when I stepped into the Bureau of Reclamation as a land-development specialist."[17]

He graduated from the University of Wyoming with a degree in agricultural economics. In 1934, amid drought and the Great Depression, his first job, as a county employee in Wyoming, was to help ranchers build earthen dams to ensure water for livestock. He built three hundred in a single county. It was there that he embarked on his dam-building career, working with local ranchers to erect small dams on ephemeral creeks to catch storm water for their scrawny cattle.

In 1937, on the drive back from a California vacation to enjoy the Rose Parade, Dominy stopped at the Bureau of Reclamation's new showcase, Boulder Dam. The colossal structure, later to be renamed Hoover Dam, was completed in 1936, and Lake Mead, created with the dam's impounded waters, had become a major tourist attraction. He left realizing—like the thousands who tour the engineering marvel each year—that the project was magnificent.

Within a decade, he was working for the Bureau, rising to the rank of assistant commissioner by 1957. Dominy was appointed commissioner in 1959 and served under Presidents Eisenhower, Kennedy, Johnson, and Nixon. His success was the result of his passion, tenacity, and persuasiveness before Congress, which kept the money flowing for his projects.

Certainly Dominy's most prominent project was the construction of a series of massive dams and reservoirs that fulfilled the wishes of the Colorado River Compact, established by an act of Congress in 1922, between the states of Arizona, California, Colorado, Nevada, New Mexico, Utah, and Wyoming. According to the act:

> The major purposes of this compact is to provide for the equitable division and apportionment of the use of the waters of the Colorado River System; to establish the relative importance of different beneficial uses of water; to promote interstate comity; to remove causes of present and future controversies and to secure the expeditious agricultural and industrial development of the Colorado River Basin, the storage of its waters, and the protection of life and property from floods.[18]

Dominy's work was vital to ensuring that all parties participating in the compact would receive their annual apportionment of water. Immediately the environmentalists of the day vehemently opposed the project because they correctly understood that an abundant water source would increase the value of personal property, certify development, and boost capitalism.

Today there are more than twenty dams on the Colorado River and its tributaries, with the primary projects being:

- The Glen Canyon dam on the mighty Colorado River—what Dominy referred to as "his crown jewel." More than seventy stories tall, the dam stores water in beautiful Lake Powell, which straddles the state lines of Utah and Arizona. Completed in 1966, the dam generates electric power for millions of residents in Utah, Arizona, and Nevada. In 1972, the popular Glen Canyon National Recreation Area was established, managed by the National Park Service.

- The Flaming Gorge dam on the Green River, a major tributary of the Colorado River in Utah, completed in 1964. The 502-foot high dam produces 151 megawatts of hydroelectric power and has created another extremely popular recreation destination. Downstream from the dam, the Green River is well known for whitewater rafting and is noted the world over for its excellent fly-fishing.

- The Navajo dam on the San Juan River, a tributary of the Colorado River in northeastern New Mexico, about thirty-four miles east of Farmington. The river is held back by what's known as a rolled-earth fill dam, which has created an embankment 402 feet high and a crest length of 3,648 feet. Despite its location on a major river and its massive height, the dam was built solely for storage purposes and thus did not have a power plant when construction was completed in 1962. In 1983, the city of Farmington contracted with Reclamation to build a 32-megawatt hydroelectric plant at the base of the dam.

- The Blue Mesa dam on the Gunnison River has created the largest body of water in Colorado. The 390-foot dam is associated with two other reservoirs that provide wonderful outdoor opportunities for thousands annually. The associated Kokanee salmon fishery is the largest in the nation. The Blue Mesa and Navajo Dams, built primarily to function for flood control purposes, have made the general regions safely habitable.

During the 1960s, Dominy was extremely effective in convincing Congress that watering the West was good for America. Nutritionists were urging Americans to diversify their diets by including fresh fruits and vegetables—not just the canned varieties—year-round. This could be better accomplished if farmers in the West had the water for irrigation. Dominy illustrated that 95 percent of the lettuce, 70 percent of the cantaloupes, 52 percent of the sweet corn, 50 percent of the carrots, and 44 percent of the cauliflower came from land watered by Reclamation Bureau projects.[19]

Dominy also pointed out that the ten most-visited reservoirs constructed by the Bureau attracted more vacationers per year than the ten most heavily used national parks. For example, in 1967, after the construction of the Glen Canyon dam, which created Lake Mead, Dominy noted that in the previous year more than 4 million "visitor days" had been spent on Lake Mead, while only 2.5 million were spent at the most heavily visited national park, Grand Teton. Today's figures are even more impressive; Lake Mead receives 10 million visitors each year, while Grand Teton hosts 3.8 million (and the adjacent Yellowstone National Park sees 2.8 million).

Dominy also understood that irrigating land increases its value, along with the value of the crops, which, in turn, creates wealth, expands the tax base of communities, and improves the quality of their schools and other public services. "The income tax increases as a result of our project growth is greater each year than the total investment in reclamation," he stated.[20]

Above all, Dominy warned that the nation had to prepare to feed a much larger population. Following World War II, the nation's population was growing exponentially, with most of the expansion occurring in the West. Meanwhile, millions of acres of marginal farmland in the South and Midwest were being retired from production. Dominy rightly forecasted the obvious: in order to be sustainable (in the true definition of the word), America needed water for agriculture, industry, and general consumption. He saw his position at the Bureau of Reclamation to be one of great responsibility and service to America.

"He relished the power he had and used it to do what he thought was the right thing to do for the country," said Roger Patterson, a former reclamation official who is now assistant general manager of the Metropolitan Water District of Southern California, in an interview with the *Los Angeles Times* following Dominy's death in 2010 at age one hundred. "He was a legend."[21]

Of course, the environmentalists disagree. Bruce Hamilton, an executive director with the Sierra Club, said, "He was a man who was promoting reclamation at a time of big dam building when people didn't understand the environmental cost of big dams nor the spiritual and societal value of free-flowing rivers."

In a conversation logged by John McPhee in his 1971 book, *Encounters with the Archdruid*, Dominy was arguing with Sierra Club chieftain, David Brower, who was lamenting the construction of the Glen Canyon dam and the subsequent creation of Lake Powell.

"Lake Powell is a drag strip for power boats... The magic of Glen Canyon is dead. It has been vulgarized."

Replied Dominy: "I'm a greater conservationist than you are, by far. I do things. I make things available to man. Unregulated, the Colorado River wouldn't be worth a good God damn to anybody."

Floyd Dominy left a legacy that enabled the Southwestern states to properly handle a growing population. Because of his tireless work, flooding is now controlled, cropland irrigation is a reality, electricity is being generated, and water for consumption is available.

By the way, only one-fifth of the Colorado River's potential hydroelectric energy has been developed, meaning that Dominy's work is yet unfinished.[22]

CUTTING FLOYD OFF

Another major water project critical to our nation's food supply, and overseen by Floyd Dominy, was construction of the Trinity River dams in

Northern California, which provide invaluable water—and electricity—to the agricultural community in California's Great Valley. The dams in question, the Trinity and Lewiston, were completed in 1963, creating one of the largest bodies of water in California. With 90 percent of the Trinity River basin's runoff able to be stored and diverted for use in other parts of the state, these dams are a key component of the Central Valley [Water] Project (CVP). Through a brilliant series of dams, tunnels, canals, and reservoirs, the CVP harnesses waters from the Trinity, Sacramento, and San Joaquin River watersheds to irrigate farmland throughout California's enormous Great Central Valley. Besides irrigation, the vast system supplies drinking water to 30 million people and generates hydroelectricity at nine power stations.

However, for years environmentalists, ever scheming to retard development and reduce population growth in California, have been pining for ways to cut the flow of water off from these Dominy dams. Finally, in 1984, their incessant lobbying began to pay off as they convinced a Democratic-led House and a moderate Republican Senate to send a bill to President Ronald Reagan, which he signed, mandating that $35.4 million be used to "restore fish and wildlife" populations along the 112-mile-long Trinity River flowing out of the dam, to combat decreased supplies of salmon, steelhead, and other species.[23] The Trinity River Basin Fish and Wildlife Restoration Act was the first slip down the proverbial slippery slope. Besides adding artificial side channels and modifying the river's banks, the act decreased the amount of water supplied to the CVP from 90 percent of the basin's runoff to 75 percent.

President Reagan was brilliant in standing up to the Soviets, but he was sucker-punched by the eco-socialists. The assumption was the money would successfully restore the fish population and that would be the end of story. But the environmentalists knew this was just the first salvo in a series of moves yet to come.

By the way, should this book inspire you to enter the political fray and become active to the point where someday you find yourself dealing with an environmental organization, remember this: your opposition is extremely patient and cunning. In what may seem like a great compromise, they'll settle for a little now, but their eventual goal is to take it all. Such has been the case with the Trinity River Basin Fish and Wildlife

Restoration Act (TRBFW).

Next, with the ascension of Clinton and Gore to the White House in 1993, radical environmentalist Bruce Babbitt was enlisted as the secretary of the interior. Babbitt immediately placed the Trinity dams in his crosshairs as he began traveling to significant dams across the country, literally carrying a sledgehammer with him. He unabashedly called his excursions the "Sledgehammer Tour" as he encouraged the destruction of America's dams.

"America overshot the mark in our dam-building frenzy," he said in a speech to the Ecological Society of America. "For most of this century, politicians have eagerly rushed in, amidst cheering crowds, to claim credit for the construction of seventy-five thousand dams all across America. Think about that number. That means we have been building, on average, one large dam a day, every single day, since the Declaration of Independence. Many of these dams have become monuments, expected to last forever. You could say forever just got a lot shorter."[24]

Immediately Babbitt unilaterally instituted a plan to further decrease the water flowing into the CVP to 52 percent, with the goal of eventually reducing the discharge to 25 percent. Babbitt's proposal was adopted by Congress, and an amendment to the 1984 Restoration Act was passed in 1995. Proving that environmentalists are always conniving for more, a provision was also neatly tucked into the buried amendment that expanded the scope of so-called habitat restoration to include another river of major concern to environmentalists, the Klamath.[25]

The headwaters of the Klamath are in Oregon, with the river flowing south into the otherwise desert farming country of southern Oregon and Northern California. Four dams have been erected on the Klamath, and three more on the associated Lost River, all part of the Klamath Project—holding water for the irrigation of 210,000 acres of productive cropland in both states and producing electricity for 150,000 homes as well. More than 700 miles of canals and channels transport the water to the necessary interests.

The damming and diverting of water in the area began with private funding in the 1800s, with the federal government becoming involved to settle water rights disputes between Oregon and California in 1905.[26]

Babbitt's public reasoning for restoring the Klamath was that the

Chinook salmon population had declined over the past hundred years as a result of altered water quality (which they insist includes the temperature of the water) due to construction of the dams and reservoirs. The solution, according to Babbitt and current activists, is to screw the farmers and the consumers and tear down the dams.

It took nearly fifteen years, but finally, in 2010, eco-activists received the bulk of their wish via a decision known as the Klamath Basin Restoration Agreement (KBRA) that calls for the removal of four hydropower dams, starting in 2020, operated by PacifiCorp, a company owned by the arguably wealthiest liberal in the world, Warren Buffett. The pact also includes restoration programs for fisheries and migrating birds, plus renewed promises of reduced water deliveries to farmers. The entire project is estimated at $1.4 billion dollars and must be approved by Congress and the secretary of the interior, who will rely on reviews by an independent panel, federal agencies, and others to determine if the decommissioning is in the public interest. However, the first independent review panel—funded by the U.S. Fish and Wildlife Service—has not been helpful to the dam-wreckers.

Released in June 2011, the 350-page report states that the panel has "strong reservations that the KBRA, as presently described, will... achieve a substantial increase in the upper basin Chinook salmon with reasonable certainty."[27]

In other words, the long-hoped-for restoration of the Klamath is not going to increase the upstream fish population.

Additionally, the scientists cautioned that the restoration plan wasn't necessarily going to correct any water quality issues, real or imagined, either. The experts stated they were "very concerned that the magnitude of the proposed solutions may not match the scope and extent of the [presumed] water quality problem."[28]

The panel's doubt centered on a flimsy computer-generated model the environmentalists were relying on to label the dams as salmon-killers. The independent reviewers detected these flaws and the junk science used to fabricate them, saying, "There are many pieces of information we do not know about the Klamath system, and none we know with absolute certainty. The process of developing the model, trying to reproduce historical conditions...must be internally consistent."[29]

According to the report, about the only way to actually increase the salmon population within the Klamath system would be to pluck fish out of the Lower Klamath River and literally truck them over a hundred miles to the Upper Klamath.

Said the panel, "A perpetual trap-and-haul program may be needed to provide adult Chinook salmon, especially the fall run, with access to the upper basin during much of the migration period."[30]

Not only would such a program be costly, the scientists warned, but fish would likely die in the process.

In the event the environmentalists get their way and the dams do come down, the panel estimates the total population of spawning fish will increase by a mere 10 percent.[31]

Let's walk through this again. Dams that provide electricity to 150,000 households, or roughly 600,000 people,[32] will be torn down; 210,000 acres of prime farmland will experience critical water reductions; and rolling aquariums will annually drive thousands of miles to transport fish to a more "suitable" location to achieve a miniscule increase in the salmon species—and the whole shebang is based on crap science. A billion dollars will come from the federal taxpayer, $200 million from PacificCorp, and another $200 million through the sale of California bonds (which is a pipe dream, given the fact that California is financially insolvent and the state's bond rating is pathetic).

And how do the environmentalists attempt to spin these renovation fees? According to an activist group known as the Klamath Riverkeepers, the outlay of money "means new jobs in Siskiyou County and a huge cash influx to local businesses."[33]

Or, more asinine Agenda 21–style "green works" jobs.

These "new jobs in Siskiyou County" will be temporary, stimulus-like gigs associated with razing the dam. Even the guys who pilot the rolling aquariums will only work a few weeks each year.

On the other hand, building and running hydroelectric dams create real, lasting jobs associated with a vital product.

Prior to Babbitt's dam-busting crusade, the hydroelectric industry directly employed nearly 48,000 people, and their earnings totaled approximately $2.7 billion, according to DOE.[34] Another 58,000 people indirectly provided services and material needed to operate and maintain hydroelectric dams and generating facilities.[35] Those employment figures are dwindling.

Nevertheless, hydropower remains one of only a few century-old industries in the United States that remains efficient and important; American plants still generate enough power to supply 28 million households with electricity.[36] A 2009 National Hydropower Association study suggests that hydropower has the potential to add 60,000 megawatts of capacity in the next fifteen years, with an industry goal to double hydro capacity and ultimately helping to create 1.4 million potential U.S. jobs.[37]

Despite the multiple benefits of hydropower, efforts to tear down America's dams and reservoirs are well organized and backed with heaps of cash.

The Nature Conservancy, an organization reporting nearly $6 billion in assets,[38] flat-out declares, "The Nature Conservancy does not advocate the building of dams or other large water-related infrastructure projects."[39]

Confirming the hard-core socialist ideology that persists within the green culture, Greenpeace (an organization that rakes in $26 million in contributions each year[40]), articulates its hatred of dams because of the plentiful jobs associated with building them, declaring, "Exploiting hydropower would give a boost to other related industries, such as machine building, iron and steel, construction materials and so forth, thus creating development opportunities for industries."[41]

TAPPING THE OCEANS

There are 12,383 miles of coastline in the United States, including Hawaii and Alaska. But eliminate those last two states and the total coastline still measures a respectable and beneficial 6,053 miles. With all of that powerful water hugging our shores, why aren't we tapping into that obvious and incredible resource to meet our water needs?

Utilizing ocean water for consumption is accomplished through a process known as *desalination*. There are more than thirteen thousand desalination plants around the world, in 120 countries, producing more than 12 billion gallons of drinkable water each day, with the most prolific producers being countries in the Middle East (Israel, Saudi Arabia, Kuwait, the United Arab Emirates, Qatar, and Bahrain) and North Africa (Libya and Algeria).

Desalination is a rather simple process. It involves forcing seawater through a series of filters and membranes to remove sand, sediment, minerals, and salt. The concentrated brine residue is discharged back into the ocean, where it naturally dilutes.

Of course, desalination has a natural enemy—the environmentalists. As usual, their arguments begin with the emotional, claiming the water intake pumps will suck in and kill sea otters, seals, fish, and other marine life. Next, they complain about the heavy concentration of salt that would be dumped into the ocean. And finally, they launch into the global warming hype that the desalination plants require significant electricity to purify the water, and that the electricity source generally is produced by natural gas or coal.

Such eco-arguments are easily dismissed. First, there are many nuclear power plants adjacent to the ocean that pump seawater to cool the reactor's core, and mechanisms are in place to ensure that marine life is not harmed. Second, the salty brine left over from desalination quickly dilutes, as oceans are anything but stagnant; they are constantly being recharged by countless streams, rivers, glaciers, and storms of snow and rain. This continual influx of fresh water maintains a consistent salinity. Third, desalting plants require a good dose of electricity to operate; however, there are emission-free alternatives to running these plants on fossil fuels. For the past thirty years, Mexico's arid Baja peninsula has employed desalination plants powered by solar energy. Perth, Australia's large desalination facility is assisted by a wind farm. In the Middle East, the preferred power source to drive desalination is nuclear energy. While nuclear is the most economically efficient way to supply power to a water plant, solar and wind could be viable alternatives, although the cost of such energy would increase the retail price of the purified product.

In the U.S., water is relatively inexpensive and plentiful compared to most other parts of the world. However, given the imprecise nature of weather, a skyrocketing projected population growth, and subsequent increases in demand for water in arid and semiarid regions of the country, there is heightened interest in desalting water as a means to augment existing supplies. In addition, many communities are interested in desalination as a cost-effective method of meeting increasingly stringent water quality regulations.

In 2008, Connecticut-based Poseidon Resources Corporation won a bid to build a $300 million desalination plant north of San Diego, California. The facility would be one of the largest in the world, producing 50 million gallons of drinking water a day, enough to supply about 100,000 homes, or as many as 400,000 people. Poseidon plans to sell the water for about $950 per acre-foot (an acre-foot is 325,851 gallons, enough water to serve 4 people for a year). That compares with an average $700 an acre-foot that local agencies now pay for water. Given that consumers are willing to pay a dollar for a 20-ounce bottle of spring water at the convenience store (there are 128 ounces in a gallon), that means we're regularly doling out nearly $6.50 for a gallon of bottled water—or $2.1 million for an acre-foot's worth of the same. Suddenly $950 for an acre-foot of desalted water sounds pretty cheap.

At the end of the day, root arguments opposing desalination mimic those against dams, reservoirs, hydropower, nuclear power, and fossil fuels, technologies encourage development, enhance lifestyles, and further American superiority. According to the Surfrider Foundation, a group who tried unsuccessfully to stop the San Diego project: "This development of 'new water' will only serve to undermine alternatives to a sustainable water supply."[42]

To the Surfriders a "sustainable water supply" translates as us being forced to use less water, as illustrated by their well-known "Kill Your Lawn" campaign.[43] The Sierra Club hates desalination too, even complaining about the "noise pollution" generated from desal plants.[44]

In ultraliberal Marin County, California, located across the Golden Gate Bridge from San Francisco, a 2009 decision by the local water board to construct a desalination plant on the San Francisco Bay was opposed by local citizens who successfully gathered signatures for a ballot initiative that stated that no such plant would be built without direct voter approval. Supporters of the initiative claimed the intake of water from the bay was filled with toxins and the proposed plant would be incapable of eliminating such materials from the newly fashioned drinking water. While such an argument is worthy of debate, alternatives that include locating the facility adjacent to the oceanside of their county has also been met with stiff opposition.

Why?

Because the activists in Marin Country are working hard with their cohorts across the nation to carry out the mission articulated by White House Science and Technology czar, John Holdren, who said, "A massive campaign must be launched to restore a high-quality environment in North America and to de-develop the United States."[45]

And their plans are much farther along than most realize.

TWELVE

RED, WHITE, AND GREEN

THE YEAR: 2050.

It's a beautiful day. And although you've been away from home for quite a while, it feels good to be back in the city you recall being the county seat. However, something isn't quite right. There are way more people than you ever remembered, outside of a special occasion, like the Fourth of July or the annual Christmas parade. The sidewalks are packed with people, but they seem emotionless, robotic, briskly walking to and fro, probably to work or shopping. No smiles. No eye contact. No hellos.

Everyone who isn't walking either navigates his way on a bike or rides a bus. You remember the downtown buses, to be sure. In fact, you used to ride them with your mom when you were a kid. But there weren't so many of them, and packed to the gills—all of them standing room only.

Where are the cars? you think to yourself. Other than a few electrically humming subcompact pretend-cars, there's nary a real one to be found.

There are lots of enormous apartment buildings. And the fancy hotel where your best friend's wedding reception was held? The new train station.

Confused, you approach a kiosk under a sign that reads, "Federal Office of Orientation, Information Center."

After just a few seconds of conversation with a badge-wearing bureaucrat behind the counter, the government worker condescendingly comments, "It's none of my business where you've been, nor do I judge you, but obviously you've been away for some time. So, let me enlighten you."

She then launches into a rote dull lecture, explaining that the population of the United States has "swollen to over 400 million"; the country is still called "the United States" but is no longer the constitutional republic consisting of fifty states that you remembered.

Instead of states, citizens now refer to their personal location of habitation by "megaregion," ten of such across the country that are subsectored into thousands of "urban hubs." Virtually everyone in America resides within one of the hubs—the closer to the center of the hub, the more convenient. Most people in the hubs live in high-density housing units close to their jobs. For those few with a little more money, home is a zero-lot-line town house or an almost-as-expensive condo; but for most, home sweet home is a tightly packed, multistory building topped with a giant solar array. The bottom floor of just about every residential building is leased to retail stores that provide the basics for living.

"Long ago people foolishly described their homes by square footage—the bigger the better. But we've since discovered the truth—smaller is more efficient," she proudly proclaims.

You are shocked to learn that people now boast about how *small* their dwelling is, which means they are fashionably reducing the size of their carbon footprint.

"That's what it's all about—your footprint. Keep thy footprint small and thus, a good citizen of the planet," she recites.

You learn that utility rates are incredibly expensive, and that electricity and water are continually in short supply. The new 3 Rs are "reuse, reduce, and recycle."

"If you would like to fill out the forms to receive monthly assistance with your electricity bill, I can direct you to the Office of Green Living. However, there is no help available with your water bill. If you surpass your quota, fines are heavy—three abuses and it's a prison sentence."

The matter-of-fact statement shocks you, and you try to cover it up. The young woman is sharp, and yet attempts to provide some odd humor, "Hey, it's not so bad, once you get the hang of it—when it's yellow, let it mellow; when it's brown, flush it down."

You're not amused.

She picks up on your discomfort and adjusts her tack, explaining that there are still a few people within the hubs that own big, traditional homes, the kind you remember. The owners are wealthy, mostly older people who live in the antiquated suburbs, where large, single-family houses remain in what decades ago were referred to as "subdivisions." The vast majority of homes in these neighborhoods were razed long ago. The big real estate killer in the suburbs was the smart meter. Once the government began closely monitoring carbon footprints and regulating electricity usage, most people couldn't afford the high energy and water bills, couldn't afford the government-mandated green upgrades, or were just fed up with the constant hassle of having their power turned off by the utility company when it determined they were consuming too much energy.

"The smart meter was a necessary government mandate that has helped heal the planet," she loyally proclaims.

Eventually, most suburbanites ended up walking away from their mortgages because there was no market for resale. The empty houses were scraped from their foundations, basements were filled in with dirt, and the properties landscaped with native species.

"There is absolutely no need for a large home in this sensible day and age, and no one does the children thing much anymore. Couples in a long-term relationship might have one child, but most choose to have none—it's far more practical."

The children thing? you think to yourself.

She pauses for a moment, and then blandly continues her memorized script.

Humans require an abundance of resources, and those are taken at Gaia's expense."

Gaia? You're stunned.

You ask about the buses and trains and the lack of cars.

Virtually everyone in the suburbs and some living in townhomes or condos own electric, plug-in cars. They're very expensive to purchase and charge. That's why for the majority of residents, walking, biking, or public transportation are the preferred options.

If it's necessary to travel to another urban hub, one can easily rent a plug-in; however, trusting an electric car to reliably take you too far is chancy. "Although we've not yet mastered the energy storage technology, our government remains committed to develop more efficient battery technology," the information officer reassures you.

"For inter-hub travel, light rail is the most efficient choice. Connections can be made with complementary bus lines to seamlessly complete your travel plans. If you must make a visit to another megaregion, then *high*-speed rail remains your most efficient choice. Punctuality is not guaranteed, but government subsidies make it the travel mode of choice for millions of citizens."

"What about air travel?" you ask.

"Well, it certainly is an option," she notes, "but keep in mind, with carbon taxes and offsets, it's a very expensive mode of transportation primarily reserved for government workers, those on business, or the rich."

The docent's face actually softens with a slight smile as she informs you of the American land beyond the megaregions.

"Thanks to the diligent work of the Department of the Interior, the Environmental Protection Agency, and the Climate Change Adaption Task Force, over the past fifteen years, the last remaining outposts of rural America have vanished," she boasts.

You can't believe what you're hearing. Private ranches, farms, and small towns are a thing of the past. Most vacation homes and cabins in more rural communities have also disappeared. While some folks lost their property to the same challenges the suburbanites succumbed to, others could no longer win battles against the EPA involving invasive, threatened, or endangered species. And others were forced to sell their property under the jurisdiction of a national park or historical landmark. Some landowners even became fearful of the overwhelming wildlife—bears, wolves, and mountain lions—that has become more plentiful each year without any natural predators, such as man, to keep their populations in check.

The kiosk agent refers to this mass evacuation of the countryside as "de-development."

She is almost excited as she goes on about how North America has now been sectored off to accommodate four massive corridors, the "wildways," designed for *wildlife only*. Other than a few ranches owned by Hollywood stars or other very wealthy citizens, evidence of past human habitation has been completely erased from these broad regions. While humans are able to visit limited portions of these vast open spaces, the bulk of the land is restricted to all but government researchers, elite members of society, and Native Americans. Violators are subject to confrontation by gun-carrying federal agents.

Bewildered, you back away from the kiosk, turn, and glance about, hoping for a sign that you're simply having a bad dream.

Where am I?

You spot a flag, an American flag. Familiar red and white stripes. Familiar fifty stars...on a field of bl—

On a field of *green*?

The brainwashed government spokesperson notices the direction of your gaze.

"Now, don't misunderstand...we're still a very patriotic people. A lot of people fly the old Stars and Stripes from their balconies. In fact, each professional soccer game begins by singing 'America the Beautiful.'"

Immediately, the accommodating bureaucrat breaks into a proud but painful rendition of the anthem, "America, America, Gaia shed her grace on thee..."

She leans toward you, whispering, "It really is a wonderful song."

"What about our national motto?" you respond.

"In Green We Trust? What about it?"

You place your hand on your forehead and desperately try to rub yourself awake.

This can't be happening, you tell yourself. It has to be a nightmare. You'll wake up any second now.

Any second...

TAKING OUR LAND

In the appendix, I've included a timetable highlighting key events in the history of the environmental movement. The chronology begins with Karl Marx's colleague, Justus von Liebig, declaring the world's first environmental disaster, perpetrated by the guano trade; continues with Marxist scholar Vladimir Lenin, declaring all forests, waters, and minerals to be the property of the state; illustrates how modern eco-diva Rachel Carson was heavily influenced by avowed socialists and communists; includes the series of green laws instituted by President Richard Nixon, including the creation of the Environmental Protection Agency; delineates the global green-washing undertaken by the United Nations; lists Al Gore's key contributions to the movement; and reminds us of the recent decisions by the Supreme Court to treat carbon dioxide as a pollutant to be regulated by the federal government.

And now we note significant additions to the timetable, provided by President Barack Obama.

Consider the executive memorandum quietly issued by the White House on April 16, 2010, deceptively titled "America's Great Outdoors Initiative."[1]

The stated goals of the Initiative:

> Reconnect...America's rivers, waterways, landscapes of national significance, ranches, farms and forests, great parks, and coasts and beaches by...creating corridors and connectivity across these outdoor spaces, and for enhancing neighborhood parks; and determine how the federal government can best advance those priorities though public/private partnerships and locally supported conservation strategies.[2]

In presenting his unilateral order, Obama sealed the deal on what promoters of the green agenda have desired for decades: "reconnect" the land of North America, creating several sweeping "corridors" designed to provide seamless "connectivity" between millions of acres of "outdoor spaces."

Notice that "ranches, farms and forests" and "coasts and beaches" are also included in the reconnection goals. This means that private ownership of such property is in the crosshairs of this scheme. The Initiative will be advanced by "the federal government...working through public/private partnerships and locally supported conservation strategies." This simple phrase empowers environmental organizations and their deep-pocketed donors to have a seat at the table with policymakers at all levels of government.

Illustrating how significant this order is, the Obama administration claims the Great Outdoors Initiative is "the most extensive expansion of land and water conservation in more than a generation."[3]

As previously addressed, nearly one-third of all our nation's land—about 700 million acres—is owned by our federal government. Most of this land is located west of the Mississippi River and is rich with oil, natural gas, and valuable mineral deposits—which they want to lock up. The initiative will aid in that ambition, and more.

The nuts and bolts of the Great Outdoors Initiative decrees four new National Conservation Areas totaling more than 330,000 acres in Colorado, New Mexico, and Utah; designates two new National Recreation Areas encompassing more than 41,000 acres in Oregon and California; enlarges the boundaries of more than a dozen existing National Parks in Ohio, New Jersey, Alabama, Louisiana, Florida, New York, and Texas; and establishes ten new National Heritage areas in Colorado, North Dakota, Maryland, Massachusetts, New Hampshire, Mississippi, Arkansas, and Alabama.

Obama's Outdoors Initiative was the cherry atop a bill he signed into law in March 2009: the National Landscape Conservation Act. This legislation, part of an omnibus-spending bill, established a 27-million-acre collage of "treasured landscapes." The acreage was originally identified by the Clinton administration and referred to as the National Landscape Conservation System (NLCS). However it took a liberal Congress and president to get the NLCS passed into law, forever enshrining these treasured landscapes to be as significant as the Statue of Liberty.

Regarding the NLCS the Obama administration states:

> The system contains many of our Nation's most treasured landscapes, including scientific, historic and cultural resources, wilderness and wilderness study areas, wild and scenic rivers, national monuments, national conservation areas, and scenic and historic trails, among others.
>
> These lands are managed as an integral part of the larger landscape, in collaboration with the neighboring landowners and surrounding communities. The management objectives are to maintain biodiversity and promote ecological connectivity and resilience in the face of climate change.[4]

It's difficult enough to define an ecosystem, let alone a landscape—or in this case a treasured landscape. An ecosystem could be as small as a pond or as large as a continent. Subscribers to the Gaia model would argue that the entire planet is a single ecosystem. In any case, an ecosystem is essentially in the eye of the beholder, and certainly the same could be said for a landscape. Now, here's the bad news for private property owners: we're told by the Department of the Interior that these "treasured landscapes" will be "managed...in collaboration with the neighboring landowners and surrounding communities," with the stated objective of maintaining "biodiversity" and promoting "ecological connectivity."

So, should you live adjacent to one of these treasured landscapes, this law declares that if your property—or *community*—is not in compliance with the government's biological and ecological standards, your land will become their land. You may continue to technically own it—or you may be forced to relinquish it—but either way, the Department of the Interior's Bureau of Land Management (BLM) will dictate how that property can or cannot be used in the name of "biodiversity" and "ecological connectivity."

What the Great Outdoors Initiative and the National Landscape Conservation System really seek to accomplish is a shocking appropriation of private real estate in order to establish "corridors and connectivity" all across the nation.

But it doesn't stop there.

BLM'S SECRET LAND GRAB

As I alluded to in the foreword, I became aware of a secret draft document procured by Obama's Department of the Interior (DOI) detailing a twenty-five-year plan to provide corridors and connectivity to enormous swaths of land. Based on dates within the paper, it was likely produced in the fourth quarter of 2009, or early 2010—before the Great Outdoors Initiative.

Soon, I was able to obtain a copy of the entire missive, entitled:

DISCUSSION PAPER[5]

Bureau of Land Management
Treasured Landscapes

Our Vision, Our Values

The document specifically denotes 140 million acres of land, administered by the Bureau of Land Management (BLM) that the government considers "treasured." According to the discussion paper, "these landscapes captured the pioneer spirit and cultivated America's romantic ideals of the Wild West."

A closer read reveals these 140 million acres (collectively about the size of Colorado or Wyoming) are not just in the West, but scattered across the country and composed of various-sized parcels, often surrounded by private property. This discussion paper addresses plans to purchase whenever possible, or take over whenever necessary, all of the land in between the government-owned parcels.

And to accomplish this goal, the draft speaks of utilizing a bevy of ecological disguises.

States the draft:

> In order to preserve these treasured landscapes for the 21st Century, the BLM proposes to manage them not as individual parcels, but as components of larger landscapes, ecosystems, airsheds and watersheds. We now know that these large-scale ecosystems, watersheds, airsheds and migratory pathways exist and function only at their natural scales, regardless of jurisdictional boundaries.

Landscapes, ecosystems, airsheds, watersheds, and migratory pathways. Virtually every acre of North America falls under such a description.

Further illustrating their intent, the DOI working paper seeks to "pursue a program of land consolidation to address its checkerboarded [sic] land...seek to acquire properties adjacent to its current holdings, if needed to preserve ecosystem integrity, and attempt to divest itself of the scattered and low-value landholdings that it has identified for disposal."

"To achieve our Treasured Landscape objectives," the document states, "the BLM will need to enlist the aid of the administration and Congress to ensure that we possess both the legal tools and financial means to make our vision of integrated landscape-level management a reality."

Legal tools to fulfill the vision would include executive orders like the Great Outdoors Initiative, more legislation similar to the NLCS, and additional funding are a must to fulfill the mission.

In the meantime, the draft says the BLM "will rely" on "three critical management tools: the Land and Water Conservation Fund, the Federal Land Transfer Facilitation Act, and a new program of renewable energy offsets."

The Land and Water Conservation Fund is a federal piggybank that provides matching grants to states and local governments for the acquisition and development of public outdoor recreation areas and facilities.

The Federal Land Transfer Facilitation Act was created in 2000 as a mechanism for the BLM to sell highly isolated plots of land that are useless to its goal of integration, and use the money to purchase parcels that will assist in the fed's plan to purchase more valuable property. The law now makes it easier for the BLM to do what President Clinton did in 1996, when administration officials, working closely with the Southern Utah Wilderness Alliance, created the Grand Staircase Escalante National Monument without consulting the Utah congressional delegation or any state or local officials. The area beneath the Utah monument is estimated to contain 62 billion tons of clean-burning, low-sulfur coal, between 3 to 5 billion barrels of oil, and 2 to 4 trillion cubic feet of natural gas. As a result of the monument designation, all of these resources are now off-limits to development.

The third management tool, "a new program of renewable energy offsets" involves, according to the internal draft, "carbon sequestration," whereby the private owner of "a stand of trees or native grasslands" will be coerced into effectively accepting a monthly or annual government bribe to ensure that the trees will not be cut down nor a field of grass plowed by a farmer. Of course this plan comes straight from the bowels of the United Nations. In a press release issued shortly before the draft was written, the U.N. Food and Agriculture agency warned "that annual greenhouse gas emissions from farming—already accounting for 14 per cent of the world's discharge—while another 17 per cent comes from deforestation and soil degradation."[6]

The paper goes on to enumerate *millions* of acres the DOI wants to acquire and lock up. To accomplish this massive taking, the document describes a few pieces of legislation that, at the time, were in the pipeline. For example, Senators Tom Udall and Jeff Bingham of New Mexico were attempting to "protect approximately 235,980 acres of BLM-managed public land." The Democratic senators' bill, the El Rio Grande del Norte National Conservation Area Establishment Act, was originally introduced in 2009, but has yet to pass. In June 2009, an attempt to protect "up to 500,000 acres" in Colorado's Dolores River Basin was introduced by Representative John Salazar (D-CO). It, too, has yet to become law.

By the way, I'm not denying that the expanses of land noted above are beautiful regions of the country. However, you must realize the eco-socialists are carefully, persistently, and simultaneously working on a variety of fronts—not to save the environment, but to control *you*. In fact, much like Lenin's nature monuments and preserves discussed in chapter 1, the internal draft actually implies that we—humans— will not be welcome in this wished-for federal land. Treasured landscapes will exist "without the trappings of visitor centers and other man-made improvements."

The discussion paper also notes its success being tied to the Federal Land Policy and Management Act of 1976, which officially declared that the U.S. was to retain remaining lands in federal ownership and use the government's power to attain even more. Quoting directly from that law:

The Congress declares that it is the official policy of the United States that—

- Public land be retained in federal ownership.[7]

- The Secretary [of Agriculture] may exercise the power of eminent domain only if necessary to secure access to public lands.[8]

- Future use [of the public land] is projected through a land use planning process coordinated with other federal and state planning efforts.[9]

- Goals and objectives be established by law as guidelines for public land use planning.[10]

- The public lands be managed in a manner that will protect the quality of scientific, scenic, historical, ecological, environmental, air and atmospheric, water resource, and archeological values; that, where appropriate, will preserve and protect certain public lands in their natural condition; that will provide food and habitat for fish and wildlife and domestic animals.[11]

- Regulations and plans for the protection of public land areas of critical environmental concern [will] be promptly developed.[12]

Perhaps the most tyrannous aspect of the Land Policy and Management Act was the government's newfound ability to use eminent domain to confiscate and acquire private property for reasons heretofore never considered. Originally the Constitution's Fifth Amendment provided eminent domain to be considered generally for the purpose of constructing infrastructure (roads, airports, reservoirs, federal buildings, etc.) that directly served the needs of the people. This act now allowed the government to "acquire...by purchase, exchange, donation, or eminent domain, lands or interest therein," with the caveat being that the properties "so acquired are confined to as narrow a corridor as is necessary to serve such purpose [in gaining access to] the National Forest System."[13]

The 1976 Act was everything the UN's Maurice Strong had hoped for when, that same year, he told the World Conference on Human Settlements, "Public ownership of land is justified in favor of the common good, rather than to protect the interest of the already privileged."[14]

Finally, the discussion paper speaks adversely of the original U.S. government land policy of the "19th and early 20th centuries [when] the public domain passed into private ownership under public land laws that made no attempt to preserve ecosystem integrity."

Only a Marxist would dislike our Founders' original intent regarding land holdings, which was to assure that all federally held land would be distributed to the states, territories, or sold into private ownership.

WILDWAYS

Dave Foreman is a rock star in radical environmental circles. Though he shuns political labels, his ideology is clearly part socialist, communist, and anarchist. Foreman has served as a Sierra Club executive board member, worked as director of wilderness affairs for the Wilderness Society, and is the founder of the eco-terror group Earth First! Foreman is forthright regarding using the environment as a tool to demonize free markets, and, in his book *Confessions of an Eco Warrior*, praises eco-activists who are "storming the barricades of capitalism."

Once, when asked to reveal his personal goals, Foreman didn't hesitate, stating, "My three main goals would be to reduce human population to about 100 million worldwide, destroy the industrial infrastructure and see wilderness, with its full complement of species, returning throughout the world."[15]

Foreman's résumé also includes founding the Wildways Project. Wildways seeks "to create contiguous networks of natural areas."[16]

"The only hope for Earth," says Foreman, "is to withdraw huge areas as inviolate natural sanctuaries from the depredations of modern industry and technology." Foreman is willing to keep metropolitan areas for human habitation, but mandate they abide by proper environmental standards. He contends the U.S. must "identify big areas that can be restored to a semblance of natural conditions, reintroduce the Grizzly Bear and wolf and prairie grasses, and declare them off limits to modern civilization."

According to the Wildland Project's current front group, the Wildlands Network (WN), the goal is to establish four transcontinental wildlife corridors. These four massive regions "are home to our greatest North American treasures: our national parks, our scenic rivers, our most majestic mountains, our crystal clear lakes, our continental trails and our vibrant grasslands and forests."[17]

The corridor furthest in the making is the Boreal Wildway that arcs east to west across Northern Canada. Through the effective actions of an organization known as the Canadian Boreal Initiative, an agreement has been brokered between activist groups such as the Wilderness Society, Greenpeace, the Nature Conservancy, and a number of timber companies to prevent logging activities and development of other resources, including oil and natural gas, in this corridor. The Canadian government has been very accommodating in advancing the creation of this wildway, especially since less than 15 percent of the country's population resides in this region. With fewer people to complain, it's been easy to obtain buy-in from both the general population of Canada and their government. Pulling a similar feat in the U.S. requires stealth, organization, politics, and patience.

The planned corridors for the U.S. include the Eastern Wildway, connecting the Everglades in Florida to the Appalachians, the Allegheny Plateau and Blue Ridge Mountains, and north to the Acadian forests of Maritime Canada. WN perceives the Eastern corridor to be the most complex to compose due to the dense population of the region: "Ultimately, it will take collective action at all scales to bring this bold vision to fruition, from creating new conservation lands, reforming policies, and providing incentives for private land stewardship, to working with transportation agencies on wildlife bridges, incorporating smart growth into local plans, and passing new legislation to face contemporary challenges."[18]

By the way, if some of the preceding terminology is new to you, it's time to get up to speed:

- "Creating new conservation lands" means acquiring land through purchase, donation, eminent domain, condemnation, protection of species, ballot initiative, law, or by federal executive power, with the express intent of forever taking the property off the books for future development of any kind.

- "Private land stewardship" is using laws pertaining to endangered, threatened or invasive species to demand private owners tread lightly on their own property, if they're allowed to tread at all. Private land stewardship policies will include enticing landowners with lucrative cap-and-trade offers to prevent their property from being farmed, ranched, or logged.

- "Working with transportation agencies on wildlife bridges" refers to strong-arming state departments of transportation to build expensive, fully landscaped bridges over multilane highways in order to provide wild animals an unbroken pathway.

- "Incorporating smart growth into local plans" speaks of the development of urban hubs, which I will describe shortly.

- "Passing new legislation to face contemporary challenges" demands that politicians who understand the Green Agenda are elected so laws, regulations and policies can be altered and created to usher in the brave new world.

The second corridor is the Western Wildway, stretching from Mexico, through the Intermountain Region, to Western Canada and Alaska. Says WN, "Our vision is coordinated international conservation action that will protect, connect, and restore a contiguous network of private and public lands along the spine of the Rocky Mountains and associated ranges, basins, plateaus, and deserts."[19]

The Pacific Wildway reaches northward from Mexico's Baja peninsula up the West Coast, including Nevada. According to the Wildway architects, "A large portion of the Pacific landscape is held in public hands, much of it in national parks, forests, and wilderness areas. Left as is, these preserves will become further isolated from each other as more roads and cities ring the preserves."[20]

Aware of the laws designed to help them achieve the grandiose goals of corridors and connectivity, WN states, "There are hundreds of organizations working on nature protection and restoration along this Wildway, many of them specializing in regional efforts."[21]

I'm sure you've witnessed examples of how "hundreds of organizations" are furthering this element of the Green Agenda in your own community. It begins with a group of activists launching a campaign to declare a small forest forever off-limits to development. An agreement by the county planning commission, or perhaps a ballot initiative, is passed declaring the land "open space" or a preserve.

Next, a defunct farm or large plot of private land adjacent to the open space is donated to, or purchased by, a nonprofit environmental group and bequeathed to the preserve. Soon, a threatened species is discovered on someone's property bordering another sector of the open space. The property owners had plans for future development, but now it's virtually worthless. A deal is worked out with the county in exchange for a tax write-off. Meanwhile another nearby parcel of land is determined to be a blight and the county uses eminent domain powers to take control of it. Another large plot of land is subject to estate taxes following the death of the owner, and the nonprofit swoops in to convince the grieving family that donating the property is their best option.

Over a period of years, the open space grows in size, eventually reaching the edge of state-owned or federal land, at which time it's transferred to the government.

This scenario is exactly what's been happening in my backyard for decades. An environmentalist outfit called the Mid-Peninsula Open Space District went from having a small patch of land under its charge in the 1960s, to having jurisdiction over sixty thousand acres today—land that is adjacent a national park, and is planned to be part of the Pacific Wildway corridor.

By the way, as Foreman hinted, the corridors will *not* be human-friendly. Allow me to describe where we lowly *Homosapiens* will be permitted to exist.

THE UNITED MEGAREGIONS OF AMERICA

Our species will be herded into megaregions.

The leading visionary organization formulating such plans is the National Committee for America 2050, a coalition of regional planning bureaucrats, academic theorists, policymakers, environmental groups, and elected leaders. The chairmen of this group are well known among their peers for their urban planning theories; Armando Carbonell, Mark Pisano, and Robert Yaro. Members of their advisory committee include Bill Clinton's secretary of the interior, the aforementioned radical environmentalist Bruce Babbitt; Don Chen, president of Smart Growth America—a man who says he was drawn to environmental policy because of his "strong awareness" of "social justice and fairness;"[22] and *Washington Post* columnist, green champion, and conservative-basher Neal Peirce. Major funding for America 2050 comes from the left-tilting Rockefeller and Ford Foundations, and the Doris Duke Foundation (Duke is the late heir to American Tobacco founder, James Buchanan Duke; the foundation donated $246 million to environmental groups in 2011).[23]

According to America 2050, "Most of the nation's rapid population growth, and an even larger share of its economic expansion, is expected to occur in ten or more emerging megaregions: large networks of metropolitan regions, each megaregion covering thousands of square miles and located in every part of the country."[24]

And if you've ever wondered why the federal government is throwing billions of dollars at projects like high-speed rail, America 2050 explains, "One way megaregions can prepare for future population pressures is by marshalling resources to make bold investments in high-speed rail and other mobility infrastructure."[25]

Quoting from the group's signature manifesto, *America 2050: A Prospectus*, the ten megaregions are described as:

- Cascadia: linking "Seattle, Portland and Vancouver, British Columbia with high-speed rail, while protecting the area's unique and pristine environment."

- Northern California: which includes the greater San Francisco Bay Area, Sacramento, and associated appendages of the Central

Valley and Sierra foothills that will offer a "high quality of life, cultural heritage, and environmental assets."

- Southern California: stretching from Santa Barbara, through the Los Angeles Basin to San Diego, with a strip connecting Las Vegas. "This region is taking aggressive action to build infrastructure that enhances its role as a global gateway while providing opportunities for its fast-growing native-born and immigrant populations."

- Texas Triangle: concentrated around Dallas-Fort Worth, Austin, and Houston, the literature notes, "Cultural cohesion creates the potential for collaboration among the metro regions of the Triangle to address land use, transportation, and environmental concerns."

- Arizona Sun Corridor: focused on Phoenix and Tucson, metro areas that "have instituted water conservation requirements and are promoting the use of desert landscaping."

- Great Lakes: this is the most expansive megaregion, extending from Southern Wisconsin through northern Illinois, engulfing nearly all of Indiana and Ohio, and including southern Michigan and western Pennsylvania. Appendages to the region connect with St. Louis and Minneapolis via interstate highways. "The region's assets include the environmental resources and amenities of the Great Lakes and a strong research and cultural tradition tied to its leading public universities."

- Northeast: comprising the populous metro areas of Boston, New York City, Philadelphia, and the Washington, D.C., metro area. "Over the next generation, the Northeast will add 18 million new residents. This population growth will demand infrastructure investments and economic growth to accommodate these new residents while preserving quality of life."

- Piedmont Atlantic: encompassing the Raleigh-Durham, North Carolina area, and reaching south to Atlanta and northern Alabama, "The region is facing challenges associated with its growing population, such as increased traffic congestion, runaway land consumption, and inadequate infrastructure, which it hopes to address with sustainable solutions."

- Gulf Coast: extending from Louisiana eastward to the tip of the Florida Panhandle, "the region is expected to grow due to the continued in-migration [*sic*] of retirees from the Midwest."

- Florida: "The Florida megaregion is one of the fastest growing in the nation and possesses a wealth of diversity, with six of every ten new residents in the last decade coming from foreign countries... Regional strategies to protect the Everglades have preserved the natural heritage of the state."

With the human population confined to the megaregions, I believe that, in the name of Treasured Landscapes and other eco-machinations, rural America will be greatly diminished, or even vanish by 2050. For those who choose to stubbornly reside in rural areas, the environmental regulations will be inexorable, and encounters with dangerous, highly protected, wild animals—bears, wolves, wildcats, and mountain lions—will be frequent.

In the meantime, the American dream will be defined as existing in a tightly packed urban hub, and being obsessed with your carbon footprint.

And there's a bevy of legislation aiming to push us into this brave new world.

CALIFORNIA'S FOREWARNING

If the slogan "As California goes, so goes the nation" has merit, and you live in a blue-leaning state, *beware*. In 2006, Governor Arnold Schwarzenegger signed AB 32 into law, the Global Warming Solutions Act, which demands that by 2020 major industrial producers of greenhouse gases must reduce their emissions by 25 percent. Other laws soon followed in California, including:

- requiring utility companies to assure that 33 percent of their energy portfolio derives from renewable energy, such as solar and wind (nuclear does not count, nor does large-scale hydropower) by 2020[26]

- instituting a cap-and-trade system later this year[27]

- demanding that diesel engines be retrofitted or replaced[28]

- prohibiting the driver of a transit bus or other commercial motor vehicle from idling more than five minutes within 100 feet of a school.[29]

- banning common bug sprays, household cleaning products, air fresheners, paint, paint thinners, and varnishes.[30]

And it only worsened when Governor Schwarzenegger signed SB 375 into law in 2008. That legislation added a futuristic spirit to AB 32.

California has been officially subdivided into eighteen "metropolitan regions," each with their own regulating agency. SB 375 demands that these agencies develop "Sustainable Communities Strategies" for land use, development, housing, and especially transportation, in order to reduce greenhouse gases. According to the law, "the transportation sector contributes over 40 percent of the greenhouse gas emissions in the State of California...the largest contributor of greenhouse gases of any sector."[31]

The goal of SB 375 is to create a socialist fantasyland in which the majority live in tightly packed, multiunit housing, walk to the store, bike to their friend's hovel, and take mass transit everywhere else.

In a report suggesting how to properly implement the law, procured by the Center for Sustainable California at the University of California–Berkeley, all government bureaucracies must work together to "help enable regions and localities to develop 'transit villages' and 'transit corridors' as vibrant, livable neighborhoods that provide not only efficient housing and transport options, but also rich public amenities such as schools, libraries, and parks."[32]

The university report actually goes so far as to acknowledge market realities that would eventually draw those with higher incomes in the new green Golden State to naturally seek nice housing closer to the center of the transit villages. However, such a progression is considered a social injustice, and therefore, "cities run the risk of allowing neighborhoods near transit to gentrify and price out lower-income families."[33]

To avoid gentrification, SB 375 encourages altering zoning laws to demand that low-income, subsidized "affordable housing"[34] be constructed adjacent to the major transit stations. Altering zoning laws in this fashion is known in government planning circles as *transfer of development rights*, or TDR. Put simply, TDR is used to force private property owners to build in such a way that is compliant with the latest government wishes—contrary to how free markets operate. Markets determine what the best use of property is, and wise owners develop accordingly. So, a property owner in the heart of the planned transit village, desiring to build an upscale condo complex or high-end retail center, should be wished lots of luck. SB 375 will encourage planners to employ TDR to force the property owner to construct an affordable housing complex, therefore limiting the owner's profit.

As the university report boasts, "Many successful TDR programs have been implemented throughout the country mostly for preservation and restoration of natural areas, protection of hillsides, preservation of historic landmarks, protection of agricultural land, promotion of urban form, and promotion of new housing and revitalization."[35]

California's landmark SB 375 is another Marxist tool to usurp property rights. In fact, the university report confirms this, proclaiming, "Firstly, a TDR program creates increased government regulation. Secondly, by separating development rights and land ownership, landowners are submitting themselves to changes in actual property rights."[36]

NATIONAL ATTEMPTS

Nationally, the piece of federal legislation that attempts to spread the California plan to every corner of America is the Livable Communities Act (LCA), sponsored in 2009 by Democratic senators Jeff Merkel of Oregon, Michael Bennet of Colorado, New York's Charles Schumer, Hawaii's Daniel Akaka, and Chris Dodd of Connecticut (now retired). As the bill notes, "Transportation accounts for 70 percent of the oil consumed in the United States and nearly one-third of carbon emissions in the United States come from the transportation sector. Reducing the growth of the number of miles driven and providing transportation alternatives through good planning and sustainable development is a necessary part of the energy independence and climate change strategies of the United States."[37]

Quoting from the LCA, an Office of Sustainable Housing and Communities will bring together all of the resources of the federal government to:

> encourage regional planning for livable communities and the adoption of sustainable development techniques, including transit-oriented development...
>
> to provide a variety of safe, reliable transportation choices, with special emphasis on public transportation...
>
> to provide affordable, energy-efficient, and location-efficient housing choices for people of all ages, incomes, races, and ethnicities...
>
> [and] to support, revitalize...and preserve undeveloped lands; to promote economic development and competitiveness by connecting the housing and employment locations of workers, reducing traffic congestion, and providing families with access to essential services.[38]

Fortunately for America, the Living Communities Act was put on ice once its liberal cosponsors realized the political winds were shifting and their congressional majority was about to crumble.

However, shortly thereafter, in a brazen abuse of executive power, President Obama issued an order to audaciously push us closer to a planned, green nation.

RED, WHITE, AND GREEN

On October 5, 2009, Barack Hussein Obama quietly signed an executive order titled, "Federal Leadership in Environmental, Energy, and Economic Performance."[39] The stated goal of the order is to "create a clean energy economy" wherein "the federal government must lead by example."

The lengthy command addresses the establishing of numerous "Steering Committees," the installation of "Sustainability Officers" in each federal agency, and charging the newly formed "Climate Change Adaptation Task Force" to recommend how the policies and practices of federal agencies can reinforce a national strategy to halt global warming.

In addition to the new layers of bureaucracy, the decree assures that the government will,

- "reduce the use of fossil fuels by using low greenhouse gas emitting vehicles,"

- "reduce potable water consumption intensity,"
- "reduce printed paper use,"
- "implement strategies...that actively support lower-carbon commuting and travel by staff,"
- ensure "all new federal buildings...are designed to achieve zero-net-energy by 2030," and
- insist "that all new construction, major renovation, or repair and alteration of federal buildings complies with" the most stringent environmental standards.

While this is to be expected from an executive command designed to ensure that the government walks the talk on all things green, there are other eco-mandates within this document that extend into the private sector. For example, "vendors and contractors" doing business, including "manufacturing, utility of delivery services, modes of transportation used, or other changes in supply chain activities," will be given "incentives to reduce greenhouse gas emissions."

Incentives would likely include gaining carbon credits should we move to a national cap-and-trade system. In the short term, an allurement would simply mean maintaining a prescribed green checklist to guarantee eligibility to bid on a government contract. So if you're a big-rig driver hauling supplies for Uncle Sam, it's likely that diesel engine of yours will need to be upgraded if you want to be hired for the job.

But that's just the start. Other recommendations include:

- "requiring vendors and contractors to register with a [yet to be named] organization for reporting greenhouse gas emissions";
- "requiring contractors...to develop and make available its greenhouse gas inventory and description of efforts to mitigate greenhouse gas emissions"; and
- "using federal government purchasing preferences or other incentives" to make sure suppliers are "using processes that minimize greenhouse gas emissions."

This is federally sponsored, green-shaded blackmail. Government vendors and contractors sell goods or services to all levels of government. They include everything from office supplies and computer equipment to consulting and network services, janitorial supplies, landscaping and construction, transportation, and utilities. This executive decision will mean that in order to do business with the U.S. government, such companies would have to prove that every aspect of their business is marked with obvious efforts to reduce their carbon footprint.

The order also empowers the feds to "advance regional and local integrated planning by

- "participating in regional transportation planning"
- "aligning federal policies to increase the effectiveness of local planning"
- "ensuring that planning for federal facilities...are pedestrian friendly, near existing employment centers, and accessible to public transit."

The centralized development of urban hubs is furthered in Section 10 of the order: "Recommendations for Sustainable Locations for Federal Facilities." Here we learn that simply the presence of a federally owned building will allow for the employment of "principles of sustainable development, including prioritizing central business districts and rural town center locations, prioritizing sites well served [*sic*] by transit...convenient pedestrian access, and consideration of transit access and proximity to housing affordable to a wide-range of federal employees." Section 10 also protects "development of sensitive land resources," meaning the feds will ramp up their demonization of private property owners, declaring certain real estate off-limits to future development.

And illustrating Team Obama's commitment to the green agenda, Section 16 pledges that the Climate Change Adaptation Task Force "is already engaged in developing the domestic and international dimensions of a U.S. strategy for adaptation to climate change."

I know what you're thinking: *This can't be happening. It has to be a nightmare. I'll wake up any second now.*

And you will. And *we* will.

AFTERWORD

IF WE DON'T QUICKLY CHANGE America's course back to our founding principles, the 2050 scenario presented in the preceding chapter could well become a reality. Preventing such a future, though, is going to require earnestly educating our fellow citizens about the treacherous road we are traveling, coupled with bold political action.

While the left is engaged in their mob-based "Occupy" protests, patriots cut from the cloth of the United States' founders must be ready to wage battle based on our country's original precepts of Life, Liberty, and the Pursuit of Happiness, as expressed in a republican form of government.

But I'm not a member of the Republican party, some might be thinking.

I'm not referring to parties; I'm differentiating between republican and democratic systems of governance.

In a republic, the general population elects representatives who are then tasked with passing laws that are based only on our Constitution to govern the nation.

By contrast, "democracy" means "rule by the people." Thus, in a democratic system, the majority rules—period—good or bad.

Many of America's founders, including James Madison, warned us of the tyranny associated with a democracy:

> "[D]emocracies have ever been spectacles of turbulence and contention; have ever been found incompatible with personal security, or the rights of property; and have, in general, been as short in their lives as they have been violent in their deaths."[1]

Unfurling the flag of eco-tyranny, the left is engaging in a devious two-fold process. On the one hand, they are deploying an aggressive, eco-based propaganda campaign designed to convince the majority of Americans that mankind's presence on Earth has irreparably harmed the planet, and the only way to prevent the environment from destruction is to create laws that will dramatically reduce greenhouse gas emissions and protect all manner of species. As we have learned, though, the spokespersons for this movement deliberately use junk science, blatant lies, propagandized school curricula, and religious institutions to sell their radical rubbish. If they are able to successfully brainwash the majority of the people, it will be all the easier to impose the 2050 scenario.

On the other hand, they are also taking advantage—and abusing—our republican form of government by doing virtually anything to elect amicable-appearing politicians, who will enter office with the subversive green agenda tucked in their briefcases. No sooner are they sworn in, these operatives begin chipping away at the Constitution, curtailing capitalism, and instituting a materialist worldview. In addition, these Marx-inspired lawmakers use all the levers of government to grow publically-funded bureaucracies, in which unelected functionaries set forth regulations and policies that strangle our personal liberties.

THE 12-POINT PLAN

Now more than ever we need selfless leaders in Washington, D.C., who are committed to the original intent of the patriots who founded this great land. Resolute representatives must be willing to immediately abolish many of the environmentally-based laws that have been described throughout this book, which, in turn, will restore liberty, reduce our budget deficit, and greatly improve the economy. Yes, such a plan would cause the left to erupt in protests, riots, and political hatchet jobs the likes of which this country has never seen. However, it is imperative that this be accomplished—for the sake of our children, our grandchildren, and *our republic*.

I believe that congress, with the consent of a conservative president, must address twelve particular points:

Abolish the Environmental Protection Agency. The EPA began with an executive order, and should end with one. The agency currently absorbs $11 billion annually. Of its 17,000 employees, a fraction could remain on the federal payroll as clean-up specialists in the event of an environmental crisis caused by the federal government, such as the crash of a future space shuttle-type craft, a domestic military accident, or overseeing the clean-up of an oil spill in federal waters.

Repeal the Federal Land Policy and Management Act of 1976 and return to the original U.S. policy of divesting of all federally-owned land. Likewise, repeal the Federal Land Transfer Facilitation Act of 2000 to insure the federal government does not have the authority to purchase additional property.

Bequeath all national parks, forests, preserves, reserves, monuments, and landscapes to the states in which they are located. Exceptions will be places with significant historical importance, such as the patriotic landmarks, memorials and museums in Washington, D.C., the Statue of Liberty, Revolutionary and Civil War Battlefields, etc. These iconic locations will be managed by a downsized National Park Service.

With the divestiture of public land, the Department of Interior could be abolished and its $19 billion annual budget erased from the bottom line. The Interior's Bureaus of Land Management, Ocean Energy Management, and Safety and Environment Enforcement, will be scrapped. The majority of the duties belonging to the Fish and Wildlife Service will be transferred to the states. Commercial fishing operations in federal waters will be overseen by Fish and Wildlife. The Bureau of Reclamation will be retained, returning to its original purpose of dam building. The U.S. Geological Survey would also be kept, but only to monitor earthquake fault lines.

In 2010, the Department of Interior established 12 renewable energy projects, built three of a planned eight regional climate change science centers, and designated more than 5,000 miles of Smart Grid transmission corridors on federal land. These projects could all be taken on by the various states, if they so choose. Negotiations between the states regarding transmission line corridors could be handled by utility companies and state agencies. Regarding the various oil and coal reserves currently locked up on federal land, once this property is given to the states, they can make decisions as to the harvest of these natural resources.

Abolish the Department of Energy. The bulk of this department's $26 billion dollar annual budget is dolled out in grants and guaranteed loans to fund pie-in-the-sky alternative energy projects, and for commissioning the Smart Grid. Private enterprise, including the utility companies, will fund these projects as they see viable, just as they once did.

Repeal the Endangered Species Act (ESA), which is administered by the Fish and Wildlife Service and the National Oceanic and Atmospheric Administration; allow such matters to be handled by the states. Issues involving fish and wildlife associated with interstate waterways will be dealt with by the effected states. NOAA's $6 billion budget could be slashed to $1 billion by cutting all but its weather forecasting services.

Amend the Outer Continental Shelf Lands Act of 1953 to restore its original intent, thus, ensuring that our massive offshore oil reserves are available for harvest.

Reverse all Obama administration orders to limit drilling in the Gulf of Mexico, off the Eastern Seaboard, and in the Alaskan Arctic.

Repeal the National Landscape Conservation Act, and, by executive order, reverse the Great Outdoors Initiative, thus emasculating the wildlife corridor plan.

Abolish all federal legislation regarding the Smart Meter and the Smart Grid.

Abolish the Clean Air Act. Allow the states to take up these issues. Prior to the roll out of the act in 1970, the various states were already well under way in deterring pollution, because their residents demanded it.

Create legislation allowing states to go forward with plans for nuclear power plant construction without fear of federal environmental

> watchdogs creating roadblocks. The Nuclear Regulatory Agency will continue setting national guidelines for plant operation and safety. An executive order will re-establish and encourage nuclear fuel reprocessing. Create legislation to provide a federal tax break to the people of Nevada in return for accepting Yucca Mountain as a national nuclear waste repository.

By electing excellent, committed patriots to employ this plan, we will be securing our liberty, while carving nearly $60 billion out of the federal budget immediately. By applying other suggestions in this book involving gas, oil, and nuclear development, we would also reap the benefit of the creation of millions of permanent, well-paying jobs, that would maintain our competitive superiority on the global stage.

The question is: How willing are we to do what clearly needs to be done?

PATRIOTIC ROLE MODELS

The stories of the patriots who founded America are incredibly inspirational—particularly the accounts of the 56 signers of the Declaration of Independence. All those who inscribed their names on this document knew that by doing so, they, and their families, would become targets of the British crown. As B.J. Lossing wrote in his 1848 book, *Signers of the Declaration of Independence*:[2]

> The signing of that instrument was a solemn act, and required great firmness and patriotism in those who committed it. It was treason against the home government, yet perfect allegiance to the law of right. It subjected those who signed it to the danger of an ignominious death, yet it entitled them to the profound reverence of a disenthralled people.

I often imagine the conversations that must have taken place between the signers and their wives prior to the vote to accept the Declaration on July 4, 1776. All of these men, but one, were married (Benjamin Franklin's wife had passed many years earlier). The family unit was very secure in those days, and the bonds of marriage were exceptionally strong. There were 55 exchanges that must have sounded something like this:

HUSBAND: *My vote and subsequent signature will guarantee vigorous persecution. The British and their allies might well come after you and our children. We will be despised by the crown.*

WIFE: *But if we don't proclaim our independence the children will grow up forever subservient to the King.*

HUSBAND: *We talk much of being ready to give all for this new land—our lives, our fortune, and our sacred honor. The battles ahead will not be easy.*

WIFE: *Neither will liberty. Sign it!*

There are three signers whose lives have always been of particular interest to me. The first is Richard Stockton. Stockton's grandparents came to New York about 1660, eventually settling near Princeton, New Jersey, where Richard would eventually be born in 1730.

Stockton became a highly regarded attorney, and, in 1766, embarked for London, where his legal skills were honored by the King. Upon returning home in 1768, Stockton was chosen as a member of the royal executive council of New Jersey and eventually placed on the bench of New Jersey Supreme Court. While it would have been natural for Stockton to remain a loyal, and wealthy, subject of the King, he longed for liberty and began to espouse the cause of the colonial patriots. The Provincial Congress of New Jersey elected him a delegate to the Continental Congress in 1776, where he became deeply involved in the debate for independence. On July 4, he voted for the Declaration, and, with the others, signed the document on August 2.

Soon after returning to his estate in Princeton, word came that the British army was coming through the area in pursuit of General Washington and his small band of soldiers. Aware that he was on the British hit list, Stockton and his wife, Annis, hastily gathered their children and fled to a friend's farm some 30 miles away. However, a neighbor faithful to the crown discovered Stockton's hideaway. A group of loyalists stormed the farm and captured Stockton and presented him to the royal authorities.

Stockton was jailed and treated extremely poorly, nearly dying of starvation. In time, Congress took up his cause and arranged a prisoner exchange to free him from his captors.

Upon release, Stockton was in terrible health. He was able to secure transportation to his estate in Princeton, and was shocked to find his home destroyed, his livestock slaughtered, his horses gone, and his wife and children in tatters.

Stockton never recovered. He suffered from chronic illness, depression, and eventually died in 1781, at the age of fifty-one. Annis and the children were cared for by family and friends.

Francis Lewis was born in Wales in 1713. He was orphaned at the age of five and raised by relatives. After a college education in London, he became a business apprentice and earnestly saved his money. At the age of twenty-one, he set sail for New York where he established an importing business.

In 1756, during the French and Indian War, Lewis was a special aid to the British forces, supplying them with uniforms and other critical supplies. He was on business at Fort Oswego, when a bloody battle broke out against the French aggressors. Lewis was taken prisoner and sent to France aboard a ship, cruelly housed in a wooden box.

Upon his release at the close of the war, Lewis was rewarded for his service to the crown with 5,000 acres of land in New York.

Again, while one might think that such a man would be forever loyal to Great Britain, such was not the case for Lewis. He saw how the edicts from England were strangling freedom in the colonies, and, according to Lossing, Lewis held dearly to "his republican views."

Lewis' wife Elizabeth was also a devout patriot and fervently supported her husband when he was elected a delegate to the General Congress in 1775 and signed the Declaration of Independence in Philadelphia the following year.

Once the Declaration was signed, the British placed a price upon the head of Francis Lewis. Before he was able to reach his home on Long Island, ground troops and a warship were sent to seize his wife and destroy his property.

Elizabeth watched from a balcony as a cannon ball crashed into a wall immediately next to her.

Immediately a servant shouted, "Run, Mistress, run!"[3]

Mrs. Lewis calmly replied, "Another shot is not likely to strike the same spot," and she did not budge.

The soldiers soon entered the home and destroyed all books, papers, and ruthlessly pillaged the entire property. Elizabeth Lewis was taken to New York and thrown into prison. She was not allowed a bed or a change of clothing, and given little to eat. A former family African servant discovered her location and was able to smuggle some small articles of clothing and some food to her. He also reported her whereabouts and condition to Congress. Demands were made for her better treatment, but the British were determined to make an example of Mrs. Lewis' prominence and wealth.

Finally, General Washington was able to broker a prisoner exchange, and Elizabeth was able to join her husband in Philadelphia. However, it was plain to everyone that because of her mistreatment, she was broken in health and was slowly sinking into the grave. Francis Lewis soon asked for a leave of absence from Congress to devote his whole time to his wife. She died in 1779.

Grief-stricken, Lewis retired from Congress to live with his sons. This great patriot died in 1802 at the age of 89 in New York City. He was buried in an unmarked grave in the yard of Trinity Church.

The third patriot who selflessly endured great sacrifice for the sake of freedom was a humble man named John Hart. He was a farmer and known throughout New Jersey as "Honest John Hart." Fellow signer, Benjamin Rush, described him as "a plain, honest, well-meaning Jersey farmer, with but little education, but with good sense and virtue enough to pursue the true interests of his country."

Honest John served with distinction as a Justice of the Peace, a Freeholder (the highest position in county government), and in the pre-Revolutionary legislatures of New Jersey. However, in 1765, he turned against the British authorities over the imposition of the Stamp Act.

The Stamp Act was a direct tax imposed on the colonies by the British parliament. The act was created to pay for British troops stationed in North America, and mandated that virtually every printed material imaginable be produced on stamped parchment produced in London, carrying an embossed revenue stamp. Like previous taxes, the stamp tax had to be paid in valid British currency, not in colonial paper money. The tax enraged many colonialists like Hart.

In 1774, Hart was elected to the first Continental Congress by the people of New Jersey, and signed the Declaration of Independence two years later.

Immediately, Hart's life was noted with a series of tragic losses. Shortly after signing the Declaration, he was elected to the New Jersey State Assembly and chosen its speaker. Knowing he was busy leading the state legislature, royal mercenaries raided his farm, destroyed his livestock, and terrorized his wife, Deborah. Upon learning of the raid, Hart immediately returned home to find his wife very ill.

Hart was at his wife's side as she passed away on October 8, 1776, but his grieving was interrupted by British troops searching for him. He fled into the forest, and his two youngest children ran to the home of a relative. Hart spent that winter on the run, sleeping in caves and eating very little. Once it became clear the British had vacated the area, Hart returned home.

Though he was re-elected as speaker of the assembly, most accounts state that Honest John's heart was broken. He soon became very ill and died at his home on May 11, 1779.

Richard and Annis Stockton, Francis and Elizabeth Lewis, and John and Deborah Hart took literally the words of the Declaration of Independence, which state:

And for the support of this Declaration, with a firm reliance on the protection of divine Providence, we mutually pledge to each other our Lives, our Fortunes and our sacred Honor.

HOW THEN SHALL WE LIVE?

If the Stocktons, Lewis', and Harts were with us today, what would they instruct us to do?

They might well say, "Arm yourselves with the truth and become engaged in battle for the future of our nation."

But what does that look like, for me? you may ask.

You can begin by absorbing books like this and passing them on. You can start a blog and write about these topics in ways in which the establishment-media never will. You can scour the Internet in search of research that trashes the green agenda.

Some of you will use the material in this book to challenge a high school teacher or college professor. Others will use some of this material to petition a local official or legislator in your state. Maybe you'll end up sharing this book with a left-leaning relative who finally acknowledges he's been sold a bill of goods.

And then there is my 12-point plan. I use those dozen points as a litmus test when speaking to conservative politicians and candidates.

I recommend we all work tirelessly to support excellent men and women of good character who are willing to run for office and reverse the green tide of legislation and policies that are destroying our nation. Some of you are able to support these candidates financially; some will act as boots on the ground with a campaign.

Some of you might be sensing a stirring in the heart to get out of your comfort zone and *run for office.*

There are no more excuses. There is little time. We need majorities in Congress to accomplish this great work. We also need a leader in the White House who will use the authority granted by the Constitution to make the challenging decisions necessary to turn the good ship U.S.A. around. As our success grows, the left will become unhinged and most of the media unglued, but liberty is not about popularity. It comes at great price.

I close with the words of Patrick Henry, the rhetorical backbone of the American Revolution, who said,

> If you make the citizens of this country agree to become the subjects of one great consolidated empire of America, your government will not have sufficient energy to keep them together. Such a government is incompatible with the genius of republicanism. There will be no checks, no real balances, in this government...If our descendants be worthy of the name of Americans they will preserve and hand down to their latest posterity the transactions of the present times.
>
> Patrick Henry, June 5, 1788

We must restore and preserve our nation as originally founded if we are to have anything of value to leave our posterity. It will certainly take much sacrifice and determination, but with God's help, it can be done.

ACKNOWLEDGMENTS

ONCE AGAIN, THANK YOU "H.L.," for your continued mentorship, encouragement, and brilliant suggestions.

A special word of appreciation must also be extended to my dear friend, Dan, who made sure the i's were dotted and the t's crossed.

And to you, beautiful Jamie Sue: "No way!"

APPENDIX

ECO-TYRANNY'S TIMELINE

1849 - Karl Marx and Friedrich Engels present, *The Communist Manifesto*, declaring "the abolition of private property."

1862 - Marx's colleague, Justus von Liebig, declares the world's first environmental disaster, perpetrated by the guano-trade. Von Liebig publishes an updated edition of *Organic Chemistry in its Application to Agriculture and Physiology*, wherein he fabricates an ecological argument to attack capitalism.

1867 - Marx releases *das Kapital*, in which he condemns the garnering profit from a natural resource.

1883 - Engels predicts a human-induced climate catastrophe: global cooling.

1905 - Marx's disciple, Sir Edwin Ray Lankester, writes a popular book, *Nature and Man*; characterizes humans as the "insurgent son" of Nature.

1918 - Influenced by Marx and Lankester, Russian dictator Vladimir Lenin writes, *On Land*. The missive declares all forests, waters, and minerals to be the property of the state.

1937 - Lankester's star pupil, Arthur Tansley, coins the term "ecosystem," and expresses deep concern about "the destructive human activities of the modern world."[1]

1939 - Future environmental icon, Bob Marshall, dies. In 1933, Marshall published *The People's Forests*, in which he advocates the public acquisition and administration of all U.S. forests in order to save the land from drilling, damming, logging, trails, and roads. In 1935, Marshall founded Wilderness Society. The organization's first goal was to see the federalization of millions of acres in Alaska.

1958 - Tansley's protégé, Charles Elton, pens a popular screed blaming humans for "an astonishing rain of death upon so much of the world's surface."[2]

1960 - After considerable lobbying by the Wilderness Society, President Eisenhower's Administration sets aside 8.9 million acres to form the core of what is today is the Alaska National Wildlife Refuge (ANWR), placing some of America's most prolific oil fields off limits.

1962 - Influenced by Tansley and Elton, former U.S. Department of Fish and Wildlife employee, Rachel Carson, releases bestseller, *Silent Spring*. In it she uses ecological issues to punish capitalism and technological progress. Carson also confirms four degrees of separation from Karl Marx as she indirectly quotes "a British biologist," Charles Elton, to describe "an amazing rain of death upon the surface of the earth."[3]

1968 - Paul Ehrlich's *Population Bomb* becomes a bestseller; claims modern development, pollution, and human population will force hundreds of millions of people to "starve to death in spite of any crash programs embarked upon now."[4]

1969 - The Union Oil spill off coast of Santa Barbara creates synergy for founding of Earth Day.

1970 - President Richard Nixon signs the National Environmental Policy Act, which ensures that environmental factors are weighted equally in decisions made by all federal agencies. The act mandates the use of Environmental Impact Statements in all federal decision making.

1970 - Nixon signs an amendment to the Clean Air Act, which allows the Federal Government to limit ozone, sulfur dioxide, nitrogen dioxide, carbon monoxide, lead, and respirable particulate matter emissions associated with industry, cars, and trucks.

1970 - Nixon unilaterally issues "Reorganization Plan Number Three" and thus establishes the Environmental Protection Agency.

1972 - The United Nations holds its first environmental conference in Stockholm. Conference chair Maurice Strong declares the gathering "must usher in a new era of international cooperation."[5] Nixon convinces Congress to give the U.N. Environmental Fund $40 million.

1976 - U.N.'s Strong organizes, "Habitat I: The World Conference on Human Settlements." The conference's preamble reveals the U.N.'s Marxist intentions: "Private land ownership is a principal instrument of accumulating wealth and therefore contributes to social injustice. Public control of land use is therefore indispensable."[6]

1976 - President Gerald Ford announces a temporary halt to nuclear fuel reprocessing in the United States. The executive order eventually terminates all commercial nuclear waste recycling in the U.S.

1976 - Ford signs the Federal Land Policy and Management Act of 1976, which asserts that the U.S. is now able to retain all federally owned land, and may exert it's power to attain even more.

1978 - Congress amends the Outer Continental Shelf Lands Act of 1953, originally passed to ensure that the massive coastal oil reserves would be *available* for harvest. The amendment turns the original law on its head by giving the government authority to cancel any offshore drilling lease or permit even if it was *thought* there could be "a substantial probability of significant impact on or damage to the coastal, marine, or human environment."

1980'S - Strong assembles a U.N. task force to establish priorities for the next Millennium. The ensuing Brundtland Commission fashions a new term, "sustainable development," and uses the handle as the basis to reorder the world's political, economic, and social structure.

1982 - James Lovelock publishes, *A New Look at Life on Earth*, and introduces the Gaia hypothesis, claiming Mother Earth is suffering from a malady—people.

1990 - Senator Al Gore popularizes the theory of global warming via his open letter published in the *New York Times*, "To Skeptics on Global Warming."

1991 - The National Religious Partnership for the Environment unveils a plan to bring environmental awareness into Christian church curricula.

1992 - Strong organizes the U.N.'s Earth Summit in Rio de Janeiro, Brazil, and presents to the world, Agenda 21. EPA Administrator William Reilly calls the Agenda "an ambitious, 900-page action plan for protecting the atmosphere, oceans, and other global resources."[7]

1992 - Senator Gore releases book, *Earth in the Balance*, and insists that global warming is irrefutable.

1993 - President Bill Clinton signs Executive Order 12852. The Order establishes the President's Council on Sustainable Development. Vice President Gore chairs the council, and announces the federal government "should not debate the science of global warming, but should instead focus on the implementation of national and local greenhouse gas reduction policies and activities."[8]

1993 - Clinton's Secretary of Interior, Bruce Babbitt, begins his "Sledgehammer Tour," and calls for the destruction of America's dams.

1994 - Gore makes way for the publication of *Growing Smart*, the comprehensive legislative guide to promote green policies, at all levels of government, across the nation.

1995 - Clinton signs amendment to the Trinity River Basin Fish and Wildlife Restoration Act, which eventually leads to withholding water from California farmers.

2000 - A consortium of U.S. environmental groups release their plans to develop urban hubs and megaregions for human habitation, and vast corridors for wildlife-only.

2000 - Clinton signs The Federal Land Transfer Facilitation Act, a mechanism for the Department of Interior to sell useless, isolated plots of federal land, and use the gain to purchase parcels that will aid in federal land integration for the creation of wildlife corridors.

2005 - President Bush signs Federal Energy Act of 2005, which cites, "It is the policy of the United States" to advance Smart Meter technology, and begin development of transcontinental Smart Grid.

2006 - Al Gore releases his Oscar-winning movie, *An Inconvenient Truth.*

2007 - Supreme Court rules carbon dioxide is a pollutant and may be regulated by the federal government via the Clean Air Act.

2009 - President Obama's EPA Administrator, Lisa Jackson, announces her agency is "authorized and obligated to reduce greenhouse gas pollutants under the Clean Air Act."

2009 - The Obama administration kills a plan put forward by President George W. Bush to develop vast shale oil deposits on federal land in Wyoming, Colorado, and Utah.

2009 - Obama signs the National Landscape Conservation Act. This legislation, part of an omnibus-spending bill, established a 27-million acre collage of "treasured landscapes."

2010 - A Secret draft document from Department of Interior is discovered, revealing the plan to designate millions of acres of land—both public and private—as protected "treasured landscapes."

2010 - An Executive Memorandum is quietly issued by the White House on April 16, 2010, deceptively titled, "America's Great Outdoors Initiative." The goal of the order is to create "corridors and connectivity" of natural, treasured areas, across the U.S.

2010 - The Obama administration suspends applications for exploratory drilling in the Alaskan Arctic.

2011 - President Obama's Energy Secretary Steven Chu determines that the Yucca Mountain, Nevada nuclear waste repository is not a "workable option." The move further stalls the expansion of nuclear energy in the United States.

The following document is the secret, Internal Draft, created by the Obama administration's Department of Interior, and the planned land heist to be overseen by the Bureau of Land Management.

The markings on the draft were apparently made by someone in the administration.

Internal Draft – NOT FOR RELEASE

DISCUSSION PAPER
Bureau of Land Management
Treasured Landscapes

Our Vision, Our Values

I. Introduction – BLM's Vision for Treasured Landscapes in the 21st Century

Of the 264 million acres under BLM management, some 130- to 140-million acres are worthy of consideration as treasured lands. These areas, roughly equivalent in size to Colorado and Wyoming combined, are valuable for their unspoiled beauty; the critical role they play in habitat conservation; their historical, cultural, and paleontological significance; and their importance in maintaining the proper functioning of the larger ecosystems in which they exist.

In order to preserve these treasured landscapes for the 21st Century, the BLM proposes to manage them not as individual parcels, but as components of larger landscapes, ecosystems, airsheds, and watersheds. We now know that these large-scale ecosystems, watersheds, airsheds, and migratory pathways exist and function only at their natural scales, regardless of jurisdictional boundaries. Therefore, in order to facilitate the transition from the current land management system, which is based on jurisdictional boundaries, to a modern landscape-level management system, the BLM proposes to "designate, rationalize, and manage-at-scale" its treasured landscape holdings.

Over the next 25 years, the BLM intends to: (1) finalize appropriate conservation designations and fully account for the ecosystem-services values of its lands; (2) rationalize and consolidate its fragmented landholdings; and (3) commit to planning and allocating resources and resource uses and at their natural scales, in effective coordination with other Federal, State, and Tribal governments. The BLM believes that together, the three components of this vision will allow us to utilize 21st century science to preserve our celebrated assets and guarantee that our treasured landscapes will be conserved for the enjoyment of future generations.

To achieve our Treasured Landscape objectives, the BLM will need to enlist the aid of the administration and Congress to ensure that we possess both the legal tools and financial means to make our vision of integrated landscape-level management a reality.

II. Background – BLM's Treasured Lands: Vast, Varied, and Vital

The BLM's lands include fragile ecosystems essential to rare animal and plant species, cultural resources that date back to the beginning of America's Native populations, stunning paleontological resources that increase our understanding of the natural world, breathtaking vistas and recreational areas, and nationally significant historic sites and trails that tell the story of our growing Nation.

The bulk of the BLM's existing treasured lands have been arrayed, by Presidential declaration, legislative enactment, or administrative management, into four separate management categories, together amounting to nearly 130-million acres—about half of BLM's total land portfolio:

1

00013553_00013596 ASLM-WDC-B1-00001-000036 Page 2 of 22

Internal Draft – NOT FOR RELEASE

- The National Landscape Conservation System: The BLM's transition to public land management on a landscape scale began a decade ago when then-Secretary Babbitt created America's newest, permanently protected collection of public lands – the National Landscape Conservation System (NLCS). By statute, the NLCS now consists of all BLM-managed National Monuments, National Conservation Areas, Wilderness, Wilderness Study Areas, Outstanding Natural Areas, Wild and Scenic Rivers, National Scenic and Historic Trails, and Conservation Lands within the California Desert—a set of public lands together comprising almost 27 million acres. These lands have been designated by Congress or Presidential proclamation to be specially managed to enhance their conservation values, while allowing for the continuation of certain multiple uses. The mission of the NLCS is to conserve, protect, and restore, for present and future generations, the nationally significant landscapes that have been recognized for their outstanding archaeological, geological, cultural, ecological, wilderness, recreation, and scientific values.

- Special Areas identified and designated through the land use planning process: Outside the NLCS, BLM land-use plans have designated about 75 million additional acres for the primary purpose of conservation and recreation. These areas include Areas of Critical Environmental Concern, Special Recreation Management Areas, Globally Important Bird Areas, Significant Caves, Research Natural Areas, National Natural Landmarks, and others.

- Areas that provide critical habitat for listed and sensitive wildlife and plant species: Further, the U.S. Fish and Wildlife Service has designated 25-million acres of BLM lands (not included in either of the two previous categories) as critical habitat for listed and sensitive wildlife and plant species.

- Wild Horse Preserves: Finally, as a result of the Secretary's October 2009 proposal to create a sustainable wild horse program, the BLM now has the opportunity to acquire preserves in the Midwest or East as part of the BLM's Treasured Landscapes initiative. America's iconic wild horses are powerful national symbols of the West, and adding federally owned wild-horse preserves to the BLM's Treasured Landscapes portfolio will provide an opportunity to expand appreciation of the BLM's conservation mission to new areas.

Because the BLM's vast landholdings hold such great promise for the Department's ambitious conservation objectives, the BLM's landscape-level preservation efforts should play a central role in the Department's Treasured Landscapes agenda.

III. The Vision: A Well-Managed System of Treasured and Protected Lands

The BLM believes that the successful management of its treasured landscapes over the next twenty-five years will require BLM to undertake three initiatives:

2

00013553_00049586 ASLM-WDC-B1-00991-000095 Page 8 of 22

Internal Draft – NOT FOR RELEASE

- First, so that lands are placed in appropriate management regimes and land-use decisions are well-informed, the BLM should ensure that its existing landholdings have received appropriate conservation designations and that the ecosystem-service values of its lands (including benefits such as carbon sequestration and air and water purification) have been adequately inventoried and considered.
- Second, to allow for more effective landscape-scale management, the BLM should aim to rationalize its land holdings by eliminating existing "checkerboard" land-holding patterns where possible, and by acquiring parcels adjacent to its current holdings, if important to preserve ecosystem integrity.
- Third, to ensure that BLM's specific land-use decisions are properly situated in their broader contexts, the BLM should commit to managing its consolidated and expanded landholdings at their natural scales, and to coordinating with other federal, state, and tribal land owners for the purposes of maintaining healthy wildlife populations, ecosystems, airsheds, watersheds, and riparian areas.

A. Completing Conservation Designations and Accounting for Ecosystem-Service Values: Expanding the NLCS, Designating New National Monuments, and Managing for Conservation in the Land-Use Planning Process.

The first component of BLM's treasured landscapes vision would ensure that the special lands already in BLM's ownership are managed under the appropriately protective management regime.

BLM-managed public lands include rugged mountains, wild deserts, and America's last vestiges of large, untamed landscapes. These landscapes first captured the pioneer spirit and cultivated America's romantic ideals of the Wild West. In order to expand this network of treasured lands to include the diversity of landscapes currently managed by the BLM and to protect world-class ecological and cultural resources, the BLM believes that lands especially deserving of protection should be placed in the National Landscape Conservation System; that the administration should consider designating significant and immediately threatened lands as national monuments; and that the BLM's land-use planning process should properly account for ecosystem-service values and manage for conservation values.

To that end, the BLM proposes that the Administration:

1. Support Congressional efforts to expand the NLCS legislatively through the designation of new National Monuments, National Conservation Areas, Wilderness Areas, Wild and Scenic Rivers, and Historic Trails. Designation efforts should not be focused solely in the West, but should also include areas in the rest of the country that warrant such protection.

3

00013553_00013596 ASLM-WDC-B4-00001-000035 Page 4 of 22

Internal Draft – NOT FOR RELEASE

2. Consider use of the Antiquities Act to set aside new National Monuments where there are immediate threats to nationally significant natural or cultural resources on lands deserving NLCS status. However, the BLM recognizes that public support and acceptance of preservation status is best achieved when the public has an opportunity to participate in a land-use-planning or legislative process.

3. Use the BLM's land-use planning process to manage for conservation values. This will allow the BLM to protect lands that, while ineligible for Monument designation and/or unlikely to receive legislative protection in the near term, are nevertheless worthy of conservation.

The BLM estimates that approximately 35 million acres of its current land holdings, all of which have been identified by the public as worthy of special protection, should be considered for a new and/or heightened conservation designation.

1. Supporting Congressional Expansion of the NLCS through Legislation Designating New Wilderness or National Conservation Areas, and Resolving the Status of BLM's existing Wilderness Study Areas.

The BLM believes that the Secretary's Treasured Landscape initiative would benefit greatly from resolution of long-standing issues relating to wilderness designations on BLM-managed public land. Currently, the BLM manages more than 545 areas, amounting to nearly 12.7-million acres, that it has identified as potentially appropriate for wilderness designation. There are strongly held opinions on each side as to whether these lands, now termed "Wilderness Study Areas," should ultimately designated as Wilderness or released for other uses. This contentious debate will continue until Congress makes a final determination as to the permanent status of these lands.

Further, the BLM recommends looking beyond its Wilderness Study Areas and building on the Secretary's expressed interest in the model of local leadership and cooperation exemplified in the recent congressional designations of Dominquez-Escalante National Monument in Colorado; Wilderness and Wild and Scenic rivers in Idaho; and National Conservation Areas, Wilderness, and Wild and Scenic Rivers in Utah. There are currently a number of locally driven proposals that would benefit from the Administration's support, including a proposal to designate New Mexico's Rio Grande del Norte as a National Conservation Area.

As a critical part of the BLM's conservation agenda, the BLM proposes working closely with the Administration and Congress to determine whether other public lands are suitable for management as National Conservation Areas. Attachment 3 contains a list of BLM-managed lands that may be appropriate for Wilderness or National Conservation Area designation.

2. Considering New National Monument Designations under the Antiquities Act.

4

00013553_00013596 ASLM-WDC-BJ-00001-000005 Page 5 of 22

Internal Draft – NOT FOR RELEASE

Should the legislative process not prove fruitful, or if a nationally significant natural or cultural land resource were to come under threat of imminent harm, BLM would recommend that the Administration consider using the Antiquities Act to designate new National Monuments by Presidential Proclamation.

The Antiquities Act allows the President to act quickly and decisively in defense of our natural and cultural treasures. Since President Theodore Roosevelt first used the Act, 15 Presidents have exercised their authority to protect nationally important places, including the Grand Tetons, Carlsbad Caverns, Mount Olympus (now Olympia National Park), Arizona's Petrified Forest, and Alaska's Denali.

For much of the history of the Act, lands designated as National Monuments were taken from the BLM and entrusted to the National Park Service. This changed in 1996, when President Clinton gave management jurisdiction over the Grand Staircase-Escalante National Monument to the BLM, reflecting a growing understanding that park-style management was not necessarily appropriate for all conservation lands. Fourteen of the 22 National Monuments created or expanded by President Clinton – totaling more than 4 million acres – remained under BLM stewardship. Today, these lands form the heart of the NLCS. BLM's management of these areas, without the trappings of visitor centers and other man-made improvements, has met with wide public support. These lands symbolize the American spirit, and their remoteness and solitude remain poignant reminders of a bygone era.

Attachment 4 contains a list of lands that may be appropriate for designation under the Antiquities Act.

3. Utilizing the Land-Use Planning Process to Account for Ecosystem-Service Values and to Protect Lands that are Ineligible for Monument Designation.

The final aspect of the first component of BLM's Treasured Landscapes vision recognizes that new conservation designations should not be the only – nor, perhaps, even the primary – means of managing for conservation on BLM's public lands. To that end, BLM also recommends emphasizing conservation values in its land-use planning process, with particular attention focused on two considerations: (a) accounting for the ecosystem-service values of BLM lands, and (b) the special legal context of conservation management in Wyoming and Alaska.

a. Accounting for Ecosystem-Service Values in the Land-Use Planning Process.

Consumptive or ground-disturbing uses of public lands often yield benefits that are readily quantifiable—BLM's timber-sale plans, for example, may be expected to yield a certain number of board feet, contribute to the creation of a definite number of jobs, and provide local communities with an estimable amount of economic activity. By contrast, non-consumptive and conserving uses of public lands have not historically been thought to yield values that are as readily quantifiable.

5

00013553_00019598 ASLM-WDC-B1-00001-000036 Page 6 of 22

Internal Draft – NOT FOR RELEASE

The modern interest in accounting for the "ecosystem services" value of lands aims to regress this imbalance. Better accounting for the value of public lands left in a condition closer to the land's natural state—whether measured in the amount of carbon sequestered by a stand of trees or native grasslands, by the economic value to local communities of undisturbed ecosystems, natural purification of air or water, or by the number of jobs retained as a result of recreational opportunities saved in an area closed to development—is designed to allow land-use decisionmakers to act with a fuller knowledge of the trade-offs involved in the choice to conserve an existing landscape, or permit new development.

Appreciating the conservation component of its multiple-use mission, the BLM proposes for its land-use planning process a new commitment to accounting for the value of the services provided by conserved ecosystems.

b. The Special Circumstance of Wyoming and Alaska

Last, it merits special mention that Congress has limited the President's authority to designate new national monuments in Wyoming and Alaska. Nevertheless, there are several landscapes in Wyoming and Alaska that contain critical ecological and cultural resources that deserve special attention and possibly enhanced protections.

The BLM therefore particularly proposes that the Administration use the BLM's land-use planning process to identify the management actions, including possible mineral withdrawals, necessary to protect sensitive resources in Wyoming and Alaska. The BLM also recommends that the Administration begin a dialogue with Congress to encourage the conservation of these areas.

Attachment 5 contains a list of areas that, though they are ineligible for Monument designation, merit protection.

B. Rationalizing BLM's Fragmented Lands: Consolidating the BLM's Checkboard-Pattern Landholdings and Enabling Landscape-Scale Ecosystem Management By Acquiring New Lands and Divesting Parcels Identified for Disposal

The second component of BLM's Treasured-Landscapes vision would involve changing the composition of BLM's public-lands portfolio.

The BLM manages what remains of the United States' once consolidated public domain. For much of the 19th and early 20th centuries, the public domain passed into private ownership (or management by other Federal agencies) under public land laws that made no attempt to preserve ecosystem integrity. In some cases, large swaths of land were patented to railroad companies in a checkerboard pattern. The resulting pattern of alternating land tenures creates extreme management difficulties, habitat fragmentation and, increasingly, user conflicts. In addition, there are today roughly 128,800 acres of inholdings in BLM-managed National Conservation Areas (excluding the California Desert) and 283,857 acres of inholdings in BLM-Managed National Monuments.

6

00013553_00013598 ASLM-WDC-B1-00001-000025 Page 7 of 22

Internal Draft – NOT FOR RELEASE

As the second component of its Treasured Landscape initiative, the BLM proposes to rationalize its existing public land holdings. Through consolidation of its protected land base and reduced fragmentation, the BLM will be better able to mitigate adverse impacts on wildlife habitat, recreation, vegetation, cultural resources, and other values. To that end, the BLM will (1) pursue a program of land consolidation to address its checkerboarded lands—particularly in Nevada, Oregon, California, Wyoming, and Utah, where the problem is most acute, (2) seek to acquire properties adjacent to its current holdings, if needed to preserve ecosystem integrity, and (3) attempt to divest itself of the scattered and low-value landholdings that it has identified for disposal through a land-use planning process.

To achieve these objectives, the BLM will rely on its land-exchange and land-acquisition programs and depend on the availability and vitality of three critical management tools: (1) the Land and Water Conservation Fund, (2) the Federal Land Transfer Facilitation Act; and (3) a new program of renewable energy offsets.

4. The Land and Water Conservation Fund (LWCF)

The Land and Water Conservation Fund (LWCF) provides a primary means for BLM to acquire lands worthy of long-term protection. LWCF funds may be used to acquire privately owned lands and waters that are adjacent to or within BLM special areas, including National Conservation Areas, National Historic and Scenic Trails, National Wild and Scenic River corridors, Wilderness, Areas of Critical Environmental Concern, and Special Recreation Management Areas.

The President's budget for 2010 includes a total of $419.9 million for LWCF programs, including $25 million for BLM; $65 million for USFWS; $68 million for NPS; and $12.1 million for appraisal services.

In FY 2010, the BLM requested LWCF funding to acquire 52,500 acres in 37 areas, with an estimated value of $58 million. In FY 2011, the BLM requested LWCF funding to acquire 47,100 acres in 40 areas, with an estimated value of $82 million. The outyear funding estimates for LWCF needs are based on a number of recent land acquisition opportunities. The majority of these monies would be used to acquire land for the NLCS and other BLM Treasured Landscapes. See Attachment 1 for more detailed LWCF funding information.

5. The Federal Land Transaction Facilitation Act (FTFLA)

The BLM is the lead agency for administering the Federal Land Transaction Facilitation Act (FLTFA). Prior to the passage of FLTFA, funds generated by BLM land sales were deposited directly into the U.S. Treasury, providing no direct conservation value to the nation.

FLTFA established the Federal Land Disposal Account that allows the BLM to sell lands with low conservation values to generate funds for the purchase of lands with high conservation

7

00013553_00013596 ASLM-WDC-B1-00001-000035 Page 8 of 22

values. Since passage of FLTFA, the sale of low-conservation value lands in the Western States has generated more than $108 million.

The BLM carefully prioritizes limited land acquisition funds to ensure every dollar is spent on only the most important and well-planned conservation projects. Typically, the BLM takes a phased approach to ensure that larger acquisitions can be completed despite temporal funding limitations. The BLM undertakes land acquisition projects within and/or contiguous to units of the NLCS and within and/or contiguous to Areas of Critical Environmental Concern.

The FLTFA expires on July 24, 2010; reauthorization legislation has been introduced by the House (H.R. 3339) and the Senate (S. 1787). To continue to allow the revenues generated from BLM's sales of isolated and difficult-to-manage public-land tracts to be an important source of funding for the acquisition of environmentally and culturally significant lands, BLM encourages the administration to strongly support the reauthorization of FLTFA. See Attachment 2 for more information regarding FLTFA disposals and approvals.

3. Using Proceeds from Renewable Energy Projects to Acquire and Protect Treasured Landscapes.

Last, as a greater amount of BLM lands are dedicated to renewable-energy projects, the BLM recommends that the administration work with Congress to enable BLM to dedicate a portion of renewable-energy proceeds to the protection and acquisition of treasured lands. In particular, BLM proposes that the administration work with Congress to (a) to require developers of renewable-energy projects to contribute to "mitigation funds," the proceeds of which would be made available to the BLM to protect public lands and mitigate the effects of development, and (b) mandate that a portion of any royalties collected from renewable-energy projects on public lands be made available to BLM to acquire additional conservation tracts.

C. Managing Problems and Ecosystems at Scale: Increasing the Use of Eco-Regional Assessments and Coordinating with other Federal, State, and Tribal Governments to manage Wildlife, Watersheds, Airsheds, and Ecosystems.

The final component of the BLM's Treasured-Landscapes vision emphasizes connectivity, and BLM's commitment to manage at a level appropriate to the issues under consideration.

The BLM recognizes that many problems and ecosystem considerations have a natural scale, and that its land-management decisions have ramifications beyond their immediate effect on BLM lands. Certain issues (such as the quality of air in a particular airshed, or the decline of a sage grouse populations in a particular region) may be best assessed, not within the confines of am artificial planning boundary, but on scales that are suggested by the physical or biological features of the issue (at the airshed, or regional sage-grouse habitat levels, for example). The BLM is just beginning to use and rely on a set of "eco-regional assessments" that are designed, in part, to enable the BLM to meaningfully engage with problems and ecosystems that cross planning-boundary lines. As the BLM looks to the next quarter century, it proposes to make increasing use of its eco-regional assessments tool.

8

00013553_00013596 ASLM-WDC-B1-00001-000036 Page 9 of 22

Internal Draft – NOT FOR RELEASE

Further, the BLM, with its extensive experience in public participation and land-use planning, coupled with the breadth of the public lands and their critical resources, stands ready with the expertise and the unparalleled capacity to coordinate with other Federal, State and Tribal Governments to tackle initiatives on a national (and even international) scale. In particular, ongoing global climate change has elevated the importance of nascent efforts to focus on cross-jurisdictional landscape connectivity and to create extensive wildlife-habitat corridors. The BLM hopes to participate fully in the effort, and to help lead the charge.

IV. Necessary Funding: Preliminary Cost Estimates

Implementing BLM's three-part Treasured Landscapes vision will require an increased investment of resources.

Over the past 10 years, funding to protect and manage lands with natural and cultural resources for use and enjoyment by current and future generations has generally remained flat when adjusted for inflation. At the same time, the West is rapidly urbanizing and the public and local governments increasingly view BLM-administered public lands as an extension of their backyards. BLM lands contribute significantly to the quality of life in the West. Taking into account the increased demands on the public lands, the BLM believes the full cost of managing Treasured Landscapes at a level commensurate with public expectations and need for conservation would be in the range of $2 to $4 per acre.

The BLM manages NLCS units for $2.21 per acre and the rest of the public lands at $1.70 per acre, totaling $59.7 million for NLCS units and $187.4 million for other Treasured Landscapes in 2010. The BLM has demonstrated that the NLCS model for conservation is significantly cost efficient in protecting landscape level resources. For example, the National Park Service operates and manages units comparatively for an average of $9.57 per acre management cost. At $4 per acre, existing and potential NLCS units could be funded for a total of $248 million by the year 2015. Other specially designated areas and habitat vital for species recovery within the BLM could be managed at $2 per acre, totaling $150.4 million by the year 2015.

The BLM recommends that any major funding increases be phased in over a five-year period to allow the BLM time to build capacity (e.g. hiring staff, developing partnerships, and processing requisitions) in order to accomplish the increased workload. This would include increased law enforcement and protection activities, expanded resource restoration and conservation activities, improved visitor services, information, education and interpretation and other activities needed to meet the increasing public demand and expectations for conservation. By the year 2015, funding for Treasured Landscapes would total $398.4 million. Projected funding needs for managing both the NLCS and the other lands comprising the BLM's Treasured Landscapes are presented in the table below by subactivity.

Proposed Funding for Treasured Landscapes showing incremental increases by Subactivity

9

00013553_00013596 ASLM-WDC-01-00001-000035 Page 10 of 22

Internal Draft – NOT FOR RELEASE

Soil, Water, Air	58,621,000	30,242,829	33,139,019	33,615,208	39,847,588
Range Mgmt	73,493,000	38,699,615	42,405,665	43,015,011	50,990,148
PD Forestry	10,443,000	3,326,711	3,645,292	3,697,673	4,383,235
Riparian	22,518,000	12,784,058	14,008,317	14,209,608	16,844,120
Cultural	15,631,000	8,874,128	9,723,954	9,863,682	11,692,444
Wildlife	35,447,000	17,099,906	18,737,470	19,006,717	22,530,631
Fisheries	13,640,000	5,068,860	5,554,277	5,634,089	6,678,669
T&E Species	22,112,000	12,553,561	13,755,746	13,953,409	16,540,421
Wilderness	18,221,000	21,169,980	23,197,313	23,530,646	27,893,311
Recreation	49,471,000	28,301,239	31,011,494	31,457,112	37,289,373
Planning	48,961,000	27,796,440	30,468,353	30,896,023	36,624,255
Law Enforcement	27,957,000	12,847,637	14,077,985	14,280,278	16,927,892
NMs & NCAs	28,801,000	31,754,970	34,795,970	35,295,969	41,839,967
Challenge Cost Share	9,500,000	5,393,398	5,909,895	5,994,817	7,106,277
O&C Forest Mgmt	31,584,000	14,906,780	16,334,320	16,569,036	19,640,994
O&C Reforestation	24,155,000	10,689,142	11,712,783	11,881,089	14,083,885
O&C Other Resources	37,544,000	18,290,428	20,042,002	20,329,994	24,099,248
O&C Planning	3,769,000	1,534,903	1,681,893	1,706,060	2,022,370
O&C NMs&NCAs	833,000	967,771	1,060,449	1,075,687	1,276,123
TOTAL	532,701,000	302,428,290	331,390,187	336,152,084	398,475,878

In addition to the funding increases necessary to support its proposed treasured-landscape management and planning functions, BLM has also prepared a series of discrete cost-estimates for its highest priority landholding rationalization efforts. They are presented in attachment 6.

V. Conclusion – Making the Vision Real

Today's Bureau of Land Management holds great promise and untapped opportunity to reflect the values of the American people. More than ever before, the public is communicating that it values healthy habitats, clean air and water, an improved quality of life, cleaner energy, and the BLM's role in supporting local economies. With the appropriate vision, the BLM can rededicate itself to the preservation of the irreplaceable cultural and historic resources in its charge, and to the effective management and conservation its treasured public lands. In doing so, it will honor the values of today's public and inspire the hopes of future generations.

10

00013553_00013596 ASIM-WDC-01-00001-000085 Page 11 of 22

Internal Draft – NOT FOR RELEASE

Attachment 1
Land and Water Conservation Fund (LWCF)

Historic BLM Funding from the Land and Water Conservation Fund Congressional Appropriations

FY 2000s
FY2000: $48,750,000
FY2001: $56,670,000
FY2002: $49,920,000
FY2003: $33,450,000
FY2004: $18,600,000
FY2005: $9,850,000
FY2006: $8,622,000
FY2007: $8,634,000
FY2008: $8,939,336
FY2009: $14,775,000

Future Anticipated BLM Funding Needs from the Land and Water Conservation Fund

FY 2010s
FY 2010: $25,029,000 (President's Budget Request)
FY 2011: $32,479,000 (President's Budget Request)[1]
FY 2012: $60,000,000 +
FY 2013: $75,000,000 +
FY 2014: $75,000,000 +
FY 2015: $75,000,000 +
FY 2016: $75,000,000 +
FY 2017: $75,000,000 +
FY 2018: $75,000,000 +
FY 2019: $75,000,000 +
FY 2020: $75,000,000 +

[1]Tentative

11

00013553_00013596 ASLM-WOC-B1-00001-000089 Page 12 of 22

Internal Draft – NOT FOR RELEASE

Attachment 2
The Federal Land Transaction Facilitation Act (FLTFA)

FLTFA Receipts for BLM Land Disposals

FY 2000s
FY 2001: $1,206,528.98
FY 2002: $2,343,807.37
FY 2003: $1,078,316.45
FY 2004: $15,759,647.48
FY 2005: $10,549,206.27
FY 2006: $57,468,523.49
FY 2007: $7,063,674.63
FY 2008: $11,555,099.52
FY 2009: $1,880,238.23
Total: $108,905,042.42

FLTFA Acquisition Approvals

FY 2000s	BLM	FWS	NPS	FS
FY 2007	$10,150,000	$1,780,000	$2,600,000	$3,500,000
FY 2008	$16,473,000	$0	$0	$13,086,950
FY 2009	$5,935,000	$800,000	$1,080,000	$1,665,250
FY 2010[1]	$14,335,000	$502,000	$0	$10,357,000

[1] Pending Secretarial Approval

12

00013553_00013598 ASLM-WDC-B1-00001-000035 Page 13 of 22

Internal Draft – NOT FOR RELEASE

Attachment 3
Conservation Designations:
Introduced and Expected Wilderness and National Conservation Area Legislation

El Rio Grande Del Norte, NM
On April 23rd, 2009, Senator Jeff Bingaman introduced legislation that would protect more than 300,000 acres of wild public land in northern New Mexico. The bill is cosponsored by Senator Tom Udall. S.874, the "El Rio Grande del Norte National Conservation Area Establishment Act would protect approximately 235,980 acres of BLM-managed public land. On June 17, 2009, Deputy Assistant Secretary for Land and Minerals Management Ned Farquhar testified before the Senate Energy and Natural Resources Committee in support of S.874. Creation of the National Conservation Area and Wilderness areas would give local communities a natural attraction and resources to use as part of a long-term sustainable economic development plan. The area is also the Rio Grande Migratory Flyway – one of the great migratory routes in the world. Eagles, falcons and hawks make the basalt walls of the Gorge their nesting homes.

Gold Butte, NV
Northeast of Las Vegas, Gold Butte is named for a historic mining town and tent city of 1,000 miners in the early 1900's. Gold Butte is much more than remnants of early mining. It is 360,000 acres of rugged mountains, Joshua tree and Mojave yucca forests, outcroppings of sandstone, and braided washes that turn into slot canyons. Gold Butte is important to numerous wildlife species, including desert tortoise, desert bighorn sheep, the banded Gila monster, great horned owls and a great variety of reptiles, birds and mammals. Gold Butte has abundant archaeological resources, including rock art, caves, agave roasting pits and camp sites dating back at least 3,000 years, and notable historical resources that deserve conservation, including Spanish and pioneer mining camps dating back to the 1700s. Legislation was introduced in the 110th Congress by Representative Shelley Berkley and similar legislation may be introduced later this year.

Organ Mountains and adjacent Wilderness, NM
Senator Bingaman has been working with community leaders and individuals in Doña Ana County in Southern New Mexico have endorsed a plan to protect 330,000 acres as wilderness and another 100,000 as a National Conservation Area in New Mexico's Organ Mountains. The mountains are home to a variety of grasses, mixed desert shrubs, piñon-juniper woodland, mixed mountain shrubs, and ponderosa pines. One of the steepest mountain ranges in the West, the Organ Mountains encompass extremely rugged terrain with steep-sided crevices, canyons, and spires.

John Day Basin, OR
The 500 mile-long John Day is the second longest free-flowing river in the lower 48 states. The river is prime habitat for wild steelhead, Chinook salmon, and bull and west slope cutthroat trout

13

00013553_00013596 ASLM-WDC-B1-00001-00003S Page 14 of 22

Internal Draft – NOT FOR RELEASE

(all protected under the Endangered Species Act). In March 2009, a bill sponsored by Senator Ron Wyden passed Congress that establishes 10,000 acres of new wilderness in the Spring Basin area of the John Day. There is potential for significant additional public land designations along this river, which would enhance recreational activities, as well as enhance habitat for one of the last strongholds for wild salmon and steelhead in the lower 48 states.

Dolores River Basin, CO
The Dolores River carves one of America's premier wild river canyons on the east side of the Colorado Plateau. The spectacular scenic landscape hosts remote wilderness, sheer-walled canyons, and magnificent stands of old-growth ponderosa pine. There is potential for up to 500,000 acres of protected public lands in this river basin. In July 2009, Representative John Salazar unveiled a proposal to establish more than 63,000 acres of wilderness and special management areas in the headwaters of this basin.

Hidden Gems (CO)
Support is being developed for legislation in north-central Colorado that would contain more than 400,000 acres of Forest Service and BLM wilderness. The areas are contained within Rep. Polis's (D-CO) and Rep. Salazar's (D-CO) districts.

Mojave Desert/Sand to Snow, CA
Located in the southern California Desert, this area includes desert tortoise habitat, critical wildlife corridors and pristine desert landscapes that have captured the imagination of Americans for decades. Senator Dianne Feinstein plans to introduce the California Desert Conservation and Recreation Act to create two new National Monuments and numerous wilderness designations.

San Diego County Wilderness, CA
This legislation would add additional acreage to the Beauty Mountain Wilderness Area in San Diego County, designated in the Omnibus Public Land Management Act. Legislation may be introduced this fall.

Utah Wilderness
Following the success of the Washington County, UT, legislation (as part of the Omnibus Public Land Act), several Utah counties including Emery, Grand, Wayne, Beaver, and Piute, have expressed interest in wilderness legislation. No county has developed legislation to date, but one or more may be developed in the 111th Congress.

14

00013553_00013596 ASLM-WDC-01-00001-000035 Page 15 of 22

Internal Draft – NOT FOR RELEASE

Attachment 4
Prospective Conservation Designation:
National Monument Designations under the Antiquities Act

Many nationally significant landscapes are worthy of inclusion in the NLCS. The areas listed below may be good candidates for National Monument designation under the Antiquities Act; however, further evaluations should be completed prior to any final decision, including an assessment of public and Congressional support.

San Rafael Swell, UT
Located in South-Central Utah, the San Rafael Swell is a 75 by 40 mile giant dome made of sandstone, shale and limestone – one of the most spectacular displays of geology in the country. The Swell is surrounded by canyons, gorges, mesas and buttes, and is home to eight rare plant species, desert big horns, coyotes, bobcats, cottontail rabbits, badgers, gray and kit fox, and the golden eagle. Visitors to the area can find ancient Indian rock art and explore a landscape with geographic features resembling those found on Mars.

Montana's Northern Prairie, MT
The Northern Montana Prairie contains some of the largest unplowed areas of grasslands in the world and some of best habitat regions in all the Great Plains. Unfortunately, we are losing our grasslands and northern prairies at alarming rates, and few opportunities exist to conserve grassland ecosystems and their native biota on large scales. If protected, Montana's Northern Prairie would connect more than 2.5 million acres of protected grasslands bordering Bitter Creek Wilderness Study Area and Grasslands National Park in Canada. This cross-boundary conservation unit would provide an opportunity to restore prairie wildlife and the possibility of establishing a new national bison range. This landscape conservation opportunity would require conservation easements, willing seller acquisitions, and withdrawal from the public domain.

Lesser Prairie Chicken Preserve, NM
This 58,000-acre Preserve is prime habitat for both the lesser prairie chicken and the sand dune lizard. This area of sand dunes and tall bluestem grasses is ideal habitat for both species. The Preserve contains more than 30 percent of the occupied lesser prairie chicken habitat in southeastern New Mexico. Recent monitoring of the area concluded that this habitat is in good to excellent condition. Protection of this area offers the best opportunity to avoid the necessity of listing either of these species as threatened or endangered.

Berryessa Snow Mountains, CA
The public lands of the Berryessa Snow Mountain region stretch from the lowlands of Putah Creek below Lake Berryessa, across remote stretches of Cache Creek, and up to the peaks of Goat Mountain and Snow Mountain. This vast expanse—nearly 500,000 acres in the wild heart of California's inner Coast Ranges — provides habitat and critical long-term movement corridors for many species of wildlife and an unusually rich part of the California Floristic Province, a biological hotspot of global importance.

15

00013553_00013596 ASLM-WDC-B1-00001-000035 Page 16 of 22

Internal Draft – NOT FOR RELEASE

Heart of the Great Basin, NV
The Heart of the Great Basin contains Nevada's wild heart – a globally unique assemblage of cultural, wildlife, and historical values. Here, Toiyabe, Toquima, and Monitor peaks tower to 12,000 feet. Thousands of petroglyphs and stone artifacts provide insight to the area's inhabitants from as long as 12,000 years ago. The region contains varied ecosystems including alpine tundra, rushing creeks, aspen groves, and high desert sage grouse habitat. The area is also a center of climate change scientific research, (e.g., Great Basin Pika is a keystone species for climate research), and one of North America's least appreciated wildland mosaics.

Otero Mesa, NM
Stretching over 1.2 million acres, Otero Mesa is home to more than 1,000 native wildlife species, including black-tailed prairie dogs, mountain lions, desert mule deer, and the only genetically pure herd of pronghorn antelope in New Mexico. These vast desert grasslands of Otero Mesa, once found throughout the region, have disappeared or been reduced to small patches unable to support native wildlife. Otero Mesa is one of the last remaining vestiges of grasslands – America's most endangered ecosystem.

Northwest Sonoran Desert, AZ
The Sonoran Desert is the most biologically diverse of all North American deserts. This area west of Phoenix is largely remote and undeveloped, with a high concentration of pristine desert wilderness landscapes. There is potential for up to 500,000 acres of new wilderness and National Conservation Area designations.

Owyhee Desert, OR/NV
Last year, Congress protected a significant portion of the Owyhee Canyonlands region in Idaho. However, a significant portion of the Owyhee region in Oregon and Nevada remains unprotected. The Owyhee Desert is one of the most remote areas in the continental United States, characterized by juniper covered deserts, natural arches, mountains and ancient lava flows. The many branching forks of the Owyhee River form deep, sheer-walled canyons between desert wilderness and entice river runners from around the Nation. The Owyhees are home to the world's largest herd of California bighorn sheep, elk, deer, cougar, Redband trout, sage-grouse and raptors.

Cascade-Siskiyou National Monument, CA (expansion)
In 2000, Cascade-Siskiyou National Monument was established to protect the extraordinary biodiversity and vegetation found in southwestern Oregon. Unfortunately, because of political constraints, the Monument's southern boundary was artificially established at the California State line. Therefore, the Monument does not include the ecologically important Klamath River tributaries and cuts out sections of important eco-regions from protection. Connectivity of landscapes is essential to protect and maintain healthy wildlife populations especially in the face of global climate change. In addition, this expansion could connect Cascade-Siskiyou with the proposed Siskiyou Crest National Monument. Expansions on the Oregon side may also be worth consideration.

16

00013553_00013596 ASLM-WDC-B1-00001-000035 Page 17 of 22

Internal Draft – NOT FOR RELEASE

Vermillion Basin, CO
The Vermillion Basin, located in northwest Colorado, is a rugged and wild landscape containing sweeping sagebrush basins, ancient petroglyph-filled canyons and whitewater rivers. Besides its scenic qualities, the basin is a critical migration corridor and winter ground for big game species such as elk, mule deer and pronghorn, in addition to being vital sage grouse habitat. This unique high desert basin is currently under threat of oil and gas development, which will forever alter the region.

Bodie Hills, CA
The remote Bodie Hills, located in the eastern Sierra Nevada, provide habitat for the imperiled sage grouse and the iconic pronghorn antelope, rare in California. The ghost town of Bodie State Historic Park, managed by the State of California, lies at the center of the Bodie Hills. Bodie State Historic Park is known as the best preserved ghost town in the West and receives several hundred thousand visits annually. Numerous gold mining operations have been proposed in the Bodies, and a new proposal is pending. Bodie Hills provides an opportunity to link both ecotourism and cultural tourism providing benefits to the surrounding communities.

The Modoc Plateau, CA
Tucked away in California's northeast corner, the Modoc Plateau contains some of the State's most spectacular and remote lands. This wild and largely undiscovered region features an array of natural riches: unbroken vistas, abundant wildlife, and millions of acres of intact, undisturbed landscapes. Spanning close to three million acres of public land that is laden with biological and archeological treasures, the Modoc Plateau is one of the State's most important natural landscapes. The crown jewel of these areas – the Skedaddle Mountains – covers close to a half-million acres in California and Nevada. The California portion alone is the second largest unprotected wilderness area in the state.

Cedar Mesa region, UT
For more than 12,000 years, generations of families from Paleo-Indian big game hunters to Mormon settlers traveled to the area now within southeastern Utah's Cedar Mesa region. Their stories are now buried among the area's estimated hundreds of thousands of prehistoric and historic sites. Cedar Mesa also contains thousands of largely intact cliff dwellings and open-air sites built between A.D. 750 and 1300 by later prehistoric farmers known as the Ancestral Puebloans or Anasazi.

San Juan Islands, WA
This cluster of hundreds of islands along the Nation's northern border contains a wealth of resources. The deep channels between islands and placid, reef-studded bays are home to myriad marine species and support major migratory routes for Orcas. The islands contain healthy pine and fir forests which protect a wide variety of wildlife species. The outstanding scenery and a historic lighthouse support diverse recreation opportunities. This area also supports sailing and sea kayaking opportunities that are unique in the Northwest.

17

00013553_00013596 ASLM-WDC-81-00001-000035 Page 18 of 22

Internal Draft – NOT FOR RELEASE

Attachment 5
Conservation Designations:
Areas worthy of protection that are ineligible for Monument Designation and unlikely to receive legislative protection in the near term

Bristol Bay Region, AK
Bristol Bay, located in southwest Alaska, is pristine wild country encompassing Alaska's largest lake, rugged snow-capped peaks and tundra laced with countless winding rivers. Bristol Bay has been called the world's greatest salmon fishery, home to the largest sockeye salmon fisheries and one of the largest king salmon runs in the world. The region is also home to caribou, brown and black bear, moose, sandhill cranes, and myriad migratory birds. Conservationists have expressed that Bristol Bay is threatened by proposed open pit gold mining, which would forever alter this pristine and delicate watershed, potentially exposing the salmon and trout habitat to acid mine drainage.

Teshekpuk Lake, AK
Teshekpuk Lake is a 22-mile wide lake located on the north slope of Alaska. Due to climate change and loss of habitat, Teshekpuk Lake has been called one of the most important areas for wildlife population survival in the entire Arctic. The Lake and surrounding land is both a migration and calving ground for 46,000 caribou and home to 90,000 summer geese. In addition, hundreds of species of birds migrate from six continents to spend part of the year at Teshekpuk Lake.

Red Desert, WY
The Red Desert's rich landscape offers spectacular desert structures and wildlife habitat. The Desert provides world class pronghorn and elk hunting; the area is home to the largest desert elk herd in North America and the migration path for 50,000 pronghorn antelope. Early explorers, pioneers, and Mormon settlers used the unique features in the Red Desert as landmarks to guide them Westward. The Pony Express Trail traverses the northern section of the Red Desert. One of the unique features in the Red Desert is Adobe Town, an astonishing and remote set of badlands and geologic formations. Visitors can see fossils of long-extinct mammals, reptiles and invertebrates.

18

00013553_00013596 ASLM-WDC-B1-00001-000035 Page 19 of 22

Internal Draft – NOT FOR RELEASE

Attachment 6
Cost Estimates: High Priority Land-Rationalization Efforts

(i) Checkerboard Consolidation
The BLM proposes a program of land consolidation for its checkerboarded lands, particularly in Nevada, Oregon, California, Wyoming, and Utah.

Cost estimate: The BLM estimates this initiative could be accomplished, where consistent with BLM land-use plans and in areas where there is a willing seller, over the next 10 years at an annual expenditure of approximately $5 million. Conversely, the BLM may use land exchanges or sales to dispose of lands within checkerboard areas consistent with land use plans as it attempts to meet our management goals for a specific area or region.

(ii) Alpine Triangle, CO
The Alpine Triangle contains a dramatic, high elevation, alpine tundra ecosystem unusual for BLM land. This wild area contains about 25,000 acres of patented mining claims that could be used to support backcountry cabins and second home development, which would threaten the landscape. Pursuing acquisition of environmentally sensitive lands here would help consolidate BLM land ownership in this nearly 200,000-acre block of high value conservation land.

Cost estimate: BLM estimates that there are approximately 2,400 patented mill sites and mining claims totaling roughly 25,000 acres. Recent Forest Service acquisitions of similarly situated groups of patented mining claims in the area were purchased for approximately $1,400 per acre. A 2008 formal appraisal for a BLM land exchange involving a small number of patented mining claims within the Triangle estimated the claims to be worth $1,700 per acre. Therefore using an average estimated value of $1,500 per acre, the total dollar amount to acquire the 25,000 acres would be about $37.5 million. This management area also includes some Forest Service Land; however, the BLM counted only patented mining claims that would fall under BLM jurisdiction if acquired. Careful analysis would be required because some claims are known to be contaminated, which would affect BLM's ability to acquire the properties.

(iii) Upper Missouri River, MT
This project is located from Fort Benton downstream to the Fort Peck Dam, a.k.a. "Fort to Fort," on the main stem of the Missouri river, along the Upper Missouri National Wild and Scenic River and Missouri Breaks National Monument, and including the Charles M. Russell National Wildlife Refuge. The stretch features a small number of very large privately owned ranches with river frontage, such as the PN Ranch along the Judith River and nearby ABN Ranch. Conserving these private ranches would benefit the Lewis and Clark National Historic Trail, the exceptional scenery along the area, and important wildlife habitat.

Cost estimate: Based on recent market activity, prices in the $300 per acre for raw land are common. For the 80,000 acres of inholdings, that would make the cost of acquiring the inholdings roughly $24 million. This would not include improvements such as houses and outbuildings, and would not necessarily include mineral rights or existing leases. The State of

19

00013553_00013598 ASLM-WDC-B1-00001-000035 Page 20 of 22

Internal Draft – NOT FOR RELEASE

Montana has also indicated a desire to divest itself of 39,000 acres of inholdings in the same area.

(iv) Pioneer Range, ID

Roughly 140,000 acres of private lands provide a critical nexus between low-elevation BLM land in the Craters of the Moon National Monument and high-elevation Forest Service lands in this region. Only about 7 percent of these lands have been protected from development by conservation easements to date. Local landowners are working with conservation groups in the Pioneers Conservation Alliance to protect this important landscape.

Cost Estimate: Costs per acre in the Pioneer Range area vary widely, from $1,000 to $20,000 per acre. Total costs would depend on the location of willing sellers.

(v) John Day River, OR

This initiative would consolidate BLM land of the John Day Wild and Scenic River in Oregon benefitting salmon recovery and allowing for more effective management of recreation along this highly scenic and popular river.

Cost estimate: To consolidate BLM lands within a quarter mile of the currently designated sections of the John Day Wild and Scenic River, it would cost approximately $67 million, working with willing sellers. This rough estimate does not factor in State and Forest Service ownership.

(vi) Upper Green River Valley, WY – Wyoming Range to Wind River Range

This initiative would focus on conserving large private ranches that are located at the base of the Wyoming and Wind River Ranges in the Upper Green River Valley to benefit sage grouse, big game species and the path of the pronghorn antelope.

Cost estimate: The BLM, the State of Wyoming, Conservation Fund, Jonah Interagency Office, Green River Valley Land Trust, Rocky Mountain Elk Foundation, Wyoming Wildlife Foundation, the Bridger Teton National Forest, and a host of other private/public partnerships are all working cooperatively in the area between the Wyoming Range and the Wind River Range to provide big game migratory corridors and wildlife habitat improvement through easements and landscape level improvement projects. These cooperative efforts pay big dividends to the State, Federal and private partners involved by increasing individual ownership and responsibility for projects, and decreasing cost and burden to Federal and State governments.

To acquire land, property values are variable based on location, features, access/availability of water, elevation, and real improvements. In order to have an accurate portrayal of costs, serious appraisal work would be required. With that in mind, a preliminary estimate of a private land purchase in the area may be calculated as follows:

Field Office Total (all ownerships)	1,618,140 acres
State & Private land:	397,210 acres (nearly a quarter of the field office area, from the Wyoming Range to the Wind River Mountains)

20

00013553_00013596 ASLM-WDC-B1-00001-000035 Page 21 of 22

Internal Draft – NOT FOR RELEASE

Average asking price per acre:	$6,000.00
Total:	$2,383,260,000

(vii) National Historic and Scenic Trails (multiple states)

This initiative would explore acquisition of key historic properties along National Historic Trails (NHTs). Willing land owners and local, State and Federal agencies work with the BLM on land acquisition, exchanges, sales, easements, and cooperative agreements, providing public access along missing segments of national scenic trails, and protection for critically important historic sites, segments, and settings along national historic trails.

Cost Estimate: Under the Secretary's Treasured Landscape initiative, the BLM would focus over the next 10 years on connecting critical scenic trail segments and the associated trail qualities, and properties that are key to the story of Western settlement and the associated diverse American cultures – including the Hispanic trails in the southwest, the journey of Lewis and Clark, the Nez Perce flight, emigrant travels West, Pony Express sites, and the Iditarod Trail between native Alaskan villages. An estimated $7 million per year would provide a substantial base for an aggressive willing seller program along the BLM's National Scenic and Historic Trails.

21

00013553_00013596 ASLM-WDC-81-00001-000035 Page 22 of 22

The White House
Office of the Press Secretary
For Immediate Release
April 16, 2010

Presidential Memorandum—America's Great Outdoors

MEMORANDUM FOR THE SECRETARY OF THE INTERIOR THE SECRETARY OF AGRICULTURE THE ADMINISTRATOR OF THE ENVIRONMENTAL PROTECTION AGENCY THE CHAIR OF THE COUNCIL ON ENVIRONMENTAL QUALITY

SUBJECT: A 21st Century Strategy for America's Great Outdoors

Americans are blessed with a vast and varied natural heritage. From mountains to deserts and from sea to shining sea, America's great outdoors have shaped the rugged independence and sense of community that define the American spirit. Our working landscapes, cultural sites, parks, coasts, wild lands, rivers, and streams are gifts that we have inherited from previous generations. They are the places that offer us refuge from daily demands, renew our spirits, and enhance our fondest memories, whether they are fishing with a grandchild in a favorite spot, hiking a trail with a friend, or enjoying a family picnic in a neighborhood park. They also are our farms, ranches, and forests—the working lands that have fed and sustained us for generations. Americans take pride in these places, and share a responsibility to preserve them for our children and grandchildren.

Today, however, we are losing touch with too many of the places and proud traditions that have helped to make America special. Farms, ranches, forests, and other valuable natural resources are disappearing at an alarming rate. Families are spending less time together enjoying their natural surroundings. Despite our conservation efforts, too many of our fields are becoming fragmented, too many of our rivers and streams are becoming polluted, and we are losing our connection to the parks, wild places, and open spaces we grew up with and cherish. Children, especially, are spending less time outside running and playing, fishing and hunting, and connecting to the outdoors just down the street or outside of town.

Across America, communities are uniting to protect the places they love, and developing new approaches to saving and enjoying the outdoors. They are bringing together farmers and ranchers, land trusts, recreation and conservation groups, sportsmen, community park groups, governments and industry, and people from all over the country to develop new partnerships and innovative programs to protect and restore our outdoors legacy. However, these efforts are often scattered and sometimes insufficient. The Federal Government, the Nation's largest land manager, has a responsibility to engage with these partners to help develop a conservation agenda worthy of the 21st Century. We must look to the private sector and nonprofit organizations, as well as towns, cities, and States, and the people who live and work in them, to identify the places that mean the most to Americans, and leverage the support of the Federal Government to help these community-driven efforts to succeed. Through these partnerships, we will work to connect these outdoor spaces to each other, and to reconnect Americans to them.

For these reasons, it is hereby ordered as follows:

Section 1. Establishment.

(a) There is established the America's Great Outdoors Initiative (Initiative), to be led by the Secretaries of the Interior and Agriculture, the Administrator of the Environmental Protection Agency, and the Chair of the Council on Environmental Quality (CEQ) and implemented in coordination with the agencies listed in section 2(b) of this memorandum. The Initiative may include the heads of other executive branch departments, agencies, and offices (agencies) as the President may, from time to time, designate.

(b) The goals of the Initiative shall be to:

(i) Reconnect Americans, especially children, to America's rivers and waterways, landscapes of national significance, ranches, farms and forests, great parks, and coasts and beaches by exploring a variety of efforts, including:

(A) promoting community-based recreation and conservation, including local parks, greenways, beaches, and waterways;

(B) advancing job and volunteer opportunities related to conservation and outdoor recreation; and

(C) supporting existing programs and projects that educate and engage Americans in our history, culture, and natural bounty.

(ii) Build upon State, local, private, and tribal priorities for the conservation of land, water, wildlife, historic, and cultural resources, creating corridors and connectivity across these outdoor spaces, and for enhancing neighborhood parks; and determine how the Federal Government can best advance those priorities through public private partnerships and locally supported conservation strategies.

(iii) Use science-based management practices to restore and protect our lands and waters for future generations.

Sec. 2. Functions. The functions of the Initiative shall include:

(a) Outreach. The Initiative shall conduct listening and learning sessions around the country where land and waters are being conserved and community parks are being established in innovative ways. These sessions should engage the full range of interested groups, including tribal leaders, farmers and ranchers, sportsmen, community park groups, foresters, youth groups, businesspeople, educators, State and local governments, and recreation and conservation groups. Special attention should be given to bringing young Americans into the conversation. These listening sessions will inform the reports required in subsection (c) of this section.

(b) Interagency Coordination. The following agencies shall work with the Initiative to identify existing resources and align policies and programs to achieve its goals:

(i) the Department of Defense;

(ii) the Department of Commerce;

(iii) the Department of Housing and Urban Development;

(iv) the Department of Health and Human Services;

(v) the Department of Labor;

(vi) the Department of Transportation;

(vii) the Department of Education; and

(viii) the Office of Management and Budget (OMB).

(c) Reports. The Initiative shall submit, through the Chair of the CEQ, the following reports to the President:

(i) Report on America's Great Outdoors. By November 15, 2010, the Initiative shall submit a report that includes the following:

(A) a review of successful and promising nonfederal conservation approaches;

(B) an analysis of existing Federal resources and programs that could be used to complement those approaches;

(C) proposed strategies and activities to achieve the goals of the Initiative; and

(D) an action plan to meet the goals of the Initiative.

The report should reflect the constraints in resources available in, and be consistent with, the Federal budget. It should recommend efficient and effective use of existing resources, as well as opportunities to leverage nonfederal public and private resources and nontraditional conservation programs.

(ii) Annual reports. By September 30, 2011, and September 30, 2012, the Initiative shall submit reports on its progress in implementing the action plan developed pursuant to subsection (c)(i)(D) of this section.

Sec. 3. General Provisions.

(a) This memorandum shall be implemented consistent with applicable law and subject to the availability of any necessary appropriations.

(b) This memorandum does not create any right or benefit, substantive or procedural, enforceable at law or in equity by any party against the United States, its departments, agencies, or entities, its officers, employees, or agents, or any other person.

(c) The heads of executive departments and agencies shall assist and provide information to the Initiative, consistent with applicable law, as may be necessary to carry out the functions of the Initiative. Each executive department and agency shall bear its own expenses of participating in the Initiative.

(d) Nothing in this memorandum shall be construed to impair or otherwise affect the functions of the Director of the OMB relating to budgetary, administrative, or legislative proposals.

(e) The Chair of the CEQ is authorized and directed to publish this memorandum in the Federal Register.

BARACK OBAMA

THE WHITE HOUSE

Office of the Press Secretary

For Immediate Release October 5, 2009

EXECUTIVE ORDER

\- - - - - - -

FEDERAL LEADERSHIP IN ENVIRONMENTAL, ENERGY, AND ECONOMIC PERFORMANCE

By the authority vested in me as President by the Constitution and the laws of the United States of America, and to establish an integrated strategy towards sustainability in the Federal Government and to make reduction of greenhouse gas emissions a priority for Federal agencies, it is hereby ordered as follows:

Section 1. Policy. In order to create a clean energy economy that will increase our Nation's prosperity, promote energy security, protect the interests of taxpayers, and safeguard the health of our environment, the Federal Government must lead by example. It is therefore the policy of the United States that Federal agencies shall increase energy efficiency; measure, report, and reduce their greenhouse gas emissions from direct and indirect activities; conserve and protect water resources through efficiency, reuse, and stormwater management; eliminate waste, recycle, and prevent pollution; leverage agency acquisitions to foster markets for sustainable technologies and environmentally preferable materials, products, and services; design, construct, maintain, and operate high performance sustainable buildings in sustainable locations; strengthen the vitality and livability of the communities in which Federal facilities are located; and inform Federal employees about and involve them in the achievement of these goals.

It is further the policy of the United States that to achieve these goals and support their respective missions, agencies shall prioritize actions based on a full accounting of both economic and social benefits and costs and shall drive continuous improvement by annually evaluating performance, extending or expanding projects that have net benefits, and reassessing or discontinuing under-performing projects.

Finally, it is also the policy of the United States that agencies' efforts and outcomes in implementing this order shall be transparent and that agencies shall therefore disclose results associated with the actions taken pursuant to this order on publicly available Federal websites.

Sec. 2. Goals for Agencies. In implementing the policy set forth in section 1 of this order, and preparing and implementing the Strategic Sustainability Performance Plan called for in section 8 of this order, the head of each agency shall:

(a) within 90 days of the date of this order, establish and report to the Chair of the Council on Environmental Quality (CEQ Chair) and the Director of the Office of Management and Budget (OMB Director) a percentage reduction target for agency-wide

more

(OVER)

reductions of scope 1 and 2 greenhouse gas emissions in absolute terms by fiscal year 2020, relative to a fiscal year 2008 baseline of the agency's scope 1 and 2 greenhouse gas emissions. Where appropriate, the target shall exclude direct emissions from excluded vehicles and equipment and from electric power produced and sold commercially to other parties in the course of regular business. This target shall be subject to review and approval by the CEQ Chair in consultation with the OMB Director under section 5 of this order. In establishing the target, the agency head shall consider reductions associated with:

(i) reducing energy intensity in agency buildings;

(ii) increasing agency use of renewable energy and implementing renewable energy generation projects on agency property; and

(iii) reducing the use of fossil fuels by:

(A) using low greenhouse gas emitting vehicles including alternative fuel vehicles;

(B) optimizing the number of vehicles in the agency fleet; and

(C) reducing, if the agency operates a fleet of at least 20 motor vehicles, the agency fleet's total consumption of petroleum products by a minimum of 2 percent annually through the end of fiscal year 2020, relative to a baseline of fiscal year 2005;

(b) within 240 days of the date of this order and concurrent with submission of the Strategic Sustainability Performance Plan as described in section 8 of this order, establish and report to the CEQ Chair and the OMB Director a percentage reduction target for reducing agency-wide scope 3 greenhouse gas emissions in absolute terms by fiscal year 2020, relative to a fiscal year 2008 baseline of agency scope 3 emissions. This target shall be subject to review and approval by the CEQ Chair in consultation with the OMB Director under section 5 of this order. In establishing the target, the agency head shall consider reductions associated with:

(i) pursuing opportunities with vendors and contractors to address and incorporate incentives to reduce greenhouse gas emissions (such as changes to manufacturing, utility or delivery services, modes of transportation used, or other changes in supply chain activities);

(ii) implementing strategies and accommodations for transit, travel, training, and conferencing that actively support lower-carbon commuting and travel by agency staff;

(iii) greenhouse gas emission reductions associated with pursuing other relevant goals in this section; and

(iv) developing and implementing innovative policies and practices to address scope 3 greenhouse gas emissions unique to agency operations;

more

3

(c) establish and report to the CEQ Chair and OMB Director a comprehensive inventory of absolute greenhouse gas emissions, including scope 1, scope 2, and specified scope 3 emissions (i) within 15 months of the date of this order for fiscal year 2010, and (ii) thereafter, annually at the end of January, for the preceding fiscal year.

(d) improve water use efficiency and management by:

(i) reducing potable water consumption intensity by 2 percent annually through fiscal year 2020, or 26 percent by the end of fiscal year 2020, relative to a baseline of the agency's water consumption in fiscal year 2007, by implementing water management strategies including water-efficient and low-flow fixtures and efficient cooling towers;

(ii) reducing agency industrial, landscaping, and agricultural water consumption by 2 percent annually or 20 percent by the end of fiscal year 2020 relative to a baseline of the agency's industrial, landscaping, and agricultural water consumption in fiscal year 2010;

(iii) consistent with State law, identifying, promoting, and implementing water reuse strategies that reduce potable water consumption; and

(iv) implementing and achieving the objectives identified in the stormwater management guidance referenced in section 14 of this order;

(e) promote pollution prevention and eliminate waste by:

(i) minimizing the generation of waste and pollutants through source reduction;

(ii) diverting at least 50 percent of non-hazardous solid waste, excluding construction and demolition debris, by the end of fiscal year 2015;

(iii) diverting at least 50 percent of construction and demolition materials and debris by the end of fiscal year 2015;

(iv) reducing printing paper use and acquiring uncoated printing and writing paper containing at least 30 percent postconsumer fiber;

(v) reducing and minimizing the quantity of toxic and hazardous chemicals and materials acquired, used, or disposed of;

(vi) increasing diversion of compostable and organic material from the waste stream;

(vii) implementing integrated pest management and other appropriate landscape management practices;

more

(OVER)

4

(viii) increasing agency use of acceptable alternative chemicals and processes in keeping with the agency's procurement policies;

(ix) decreasing agency use of chemicals where such decrease will assist the agency in achieving greenhouse gas emission reduction targets under section 2(a) and (b) of this order; and

(x) reporting in accordance with the requirements of sections 301 through 313 of the Emergency Planning and Community Right-to-Know Act of 1986 (42 U.S.C. 11001 *et seq.*);

(f) advance regional and local integrated planning by:

(i) participating in regional transportation planning and recognizing existing community transportation infrastructure;

(ii) aligning Federal policies to increase the effectiveness of local planning for energy choices such as locally generated renewable energy;

(iii) ensuring that planning for new Federal facilities or new leases includes consideration of sites that are pedestrian friendly, near existing employment centers, and accessible to public transit, and emphasizes existing central cities and, in rural communities, existing or planned town centers;

(iv) identifying and analyzing impacts from energy usage and alternative energy sources in all Environmental Impact Statements and Environmental Assessments for proposals for new or expanded Federal facilities under the National Environmental Policy Act of 1969, as amended (42 U.S.C. 4321 *et seq.*); and

(v) coordinating with regional programs for Federal, State, tribal, and local ecosystem, watershed, and environmental management;

(g) implement high performance sustainable Federal building design, construction, operation and management, maintenance, and deconstruction including by:

(i) beginning in 2020 and thereafter, ensuring that all new Federal buildings that enter the planning process are designed to achieve zero-net-energy by 2030;

(ii) ensuring that all new construction, major renovation, or repair and alteration of Federal buildings complies with the *Guiding Principles for Federal Leadership in High Performance and Sustainable Buildings* (Guiding Principles);

(iii) ensuring that at least 15 percent of the agency's existing buildings (above 5,000 gross square feet) and building leases (above 5,000

more

5

gross square feet) meet the Guiding Principles by fiscal year 2015 and that the agency makes annual progress toward 100-percent conformance with the Guiding Principles for its building inventory;

(iv) pursuing cost-effective, innovative strategies, such as highly reflective and vegetated roofs, to minimize consumption of energy, water, and materials;

(v) managing existing building systems to reduce the consumption of energy, water, and materials, and identifying alternatives to renovation that reduce existing assets' deferred maintenance costs;

(vi) when adding assets to the agency's real property inventory, identifying opportunities to consolidate and dispose of existing assets, optimize the performance of the agency's real-property portfolio, and reduce associated environmental impacts; and

(vii) ensuring that rehabilitation of federally owned historic buildings utilizes best practices and technologies in retrofitting to promote long-term viability of the buildings;

(h) advance sustainable acquisition to ensure that 95 percent of new contract actions including task and delivery orders, for products and services with the exception of acquisition of weapon systems, are energy-efficient (Energy Star or Federal Energy Management Program (FEMP) designated), water-efficient, biobased, environmentally preferable (e.g., Electronic Product Environmental Assessment Tool (EPEAT) certified), non-ozone depleting, contain recycled content, or are non-toxic or less-toxic alternatives, where such products and services meet agency performance requirements;

(i) promote electronics stewardship, in particular by:

(i) ensuring procurement preference for EPEAT-registered electronic products;

(ii) establishing and implementing policies to enable power management, duplex printing, and other energy-efficient or environmentally preferable features on all eligible agency electronic products;

(iii) employing environmentally sound practices with respect to the agency's disposition of all agency excess or surplus electronic products;

(iv) ensuring the procurement of Energy Star and FEMP designated electronic equipment;

(v) implementing best management practices for energy-efficient management of servers and Federal data centers; and

more

(OVER)

6

(j) sustain environmental management, including by:

(i) continuing implementation of formal environmental management systems at all appropriate organizational levels; and

(ii) ensuring these formal systems are appropriately implemented and maintained to achieve the performance necessary to meet the goals of this order.

Sec. 3. Steering Committee on Federal Sustainability. The OMB Director and the CEQ Chair shall:

(a) establish an interagency Steering Committee (Steering Committee) on Federal Sustainability composed of the Federal Environmental Executive, designated under section 6 of Executive Order 13423 of January 24, 2007, and Agency Senior Sustainability Officers, designated under section 7 of this order, and that shall:

(i) serve in the dual capacity of the Steering Committee on Strengthening Federal Environmental, Energy, and Transportation Management designated by the CEQ Chair pursuant to section 4 of Executive Order 13423;

(ii) advise the OMB Director and the CEQ Chair on implementation of this order;

(iii) facilitate the implementation of each agency's Strategic Sustainability Performance Plan; and

(iv) share information and promote progress towards the goals of this order;

(b) enlist the support of other organizations within the Federal Government to assist the Steering Committee in addressing the goals of this order;

(c) establish and disband, as appropriate, interagency subcommittees of the Steering Committee, to assist the Steering Committee in carrying out its responsibilities;

(d) determine appropriate Federal actions to achieve the policy of section 1 and the goals of section 2 of this order;

(e) ensure that Federal agencies are held accountable for conformance with the requirements of this order; and

(f) in coordination with the Department of Energy's Federal Energy Management Program and the Office of the Federal Environmental Executive designated under section 6 of Executive Order 13423, provide guidance and assistance to facilitate the development of agency targets for greenhouse gas emission reductions required under subsections 2(a) and (b) of this order.

Sec. 4. Additional Duties of the Director of the Office of Management and Budget. In addition to the duties of the OMB Director specified elsewhere in this order, the OMB Director shall:

more

7

(a) review and approve each agency's multi-year Strategic Sustainability Performance Plan under section 8 of this order and each update of the Plan. The Director shall, where feasible, review each agency's Plan concurrently with OMB's review and evaluation of the agency's budget request;

(b) prepare scorecards providing periodic evaluation of Federal agency performance in implementing this order and publish scorecard results on a publicly available website; and

(c) approve and issue instructions to the heads of agencies concerning budget and appropriations matters relating to implementation of this order.

Sec. 5. Additional Duties of the Chair of the Council on Environmental Quality. In addition to the duties of the CEQ Chair specified elsewhere in this order, the CEQ Chair shall:

(a) issue guidance for greenhouse gas accounting and reporting required under section 2 of this order;

(b) issue instructions to implement this order, in addition to instructions within the authority of the OMB Director to issue under subsection 4(c) of this order;

(c) review and approve each agency's targets, in consultation with the OMB Director, for agency-wide reductions of greenhouse gas emissions under section 2 of this order;

(d) prepare, in coordination with the OMB Director, streamlined reporting metrics to determine each agency's progress under section 2 of this order;

(e) review and evaluate each agency's multi-year Strategic Sustainability Performance Plan under section 8 of this order and each update of the Plan;

(f) assess agency progress toward achieving the goals and policies of this order, and provide its assessment of the agency's progress to the OMB Director;

(g) within 120 days of the date of this order, provide the President with an aggregate Federal Government-wide target for reducing scope 1 and 2 greenhouse gas emissions in absolute terms by fiscal year 2020 relative to a fiscal year 2008 baseline;

(h) within 270 days of the date of this order, provide the President with an aggregate Federal Government-wide target for reducing scope 3 greenhouse gas emissions in absolute terms by fiscal year 2020 relative to a fiscal year 2008 baseline;

(i) establish and disband, as appropriate, interagency working groups to provide recommendations to the CEQ for areas of Federal agency operational and managerial improvement associated with the goals of this order; and

(j) administer the Presidential leadership awards program, established under subsection 4(c) of Executive Order 13423, to recognize exceptional and outstanding agency performance with respect to achieving the goals of this order and to recognize extraordinary innovation, technologies, and practices employed to achieve the goals of this order.

more

(OVER)

8

Sec. 6. Duties of the Federal Environmental Executive. The Federal Environmental Executive designated by the President to head the Office of the Federal Environmental Executive, pursuant to section 6 of Executive Order 13423, shall:

(a) identify strategies and tools to assist Federal implementation efforts under this order, including through the sharing of best practices from successful Federal sustainability efforts; and

(b) monitor and advise the CEQ Chair and the OMB Director on the agencies' implementation of this order and their progress in achieving the order's policies and goals.

Sec. 7. Agency Senior Sustainability Officers. (a) Within 30 days of the date of this order, the head of each agency shall designate from among the agency's senior management officials a Senior Sustainability Officer who shall be accountable for agency conformance with the requirements of this order; and shall report such designation to the OMB Director and the CEQ Chair.

(b) The Senior Sustainability Officer for each agency shall perform the functions of the senior agency official designated by the head of each agency pursuant to section 3(d)(i) of Executive Order 13423 and shall be responsible for:

(i) preparing the targets for agency-wide reductions and the inventory of greenhouse gas emissions required under subsections 2(a), (b), and (c) of this order;

(ii) within 240 days of the date of this order, and annually thereafter, preparing and submitting to the CEQ Chair and the OMB Director, for their review and approval, a multi-year Strategic Sustainability Performance Plan (Sustainability Plan or Plan) as described in section 8 of this order;

(iii) preparing and implementing the approved Plan in coordination with appropriate offices and organizations within the agency including the General Counsel, Chief Information Officer, Chief Acquisition Officer, Chief Financial Officer, and Senior Real Property Officers, and in coordination with other agency plans, policies, and activities;

(iv) monitoring the agency's performance and progress in implementing the Plan, and reporting the performance and progress to the CEQ Chair and the OMB Director, on such schedule and in such format as the Chair and the Director may require; and

(v) reporting annually to the head of the agency on the adequacy and effectiveness of the agency's Plan in implementing this order.

Sec. 8. Agency Strategic Sustainability Performance Plan. Each agency shall develop, implement, and annually update an integrated Strategic Sustainability Performance Plan that will prioritize agency actions based on lifecycle return

more

9

on investment. Each agency Plan and update shall be subject to approval by the OMB Director under section 4 of this order. With respect to the period beginning in fiscal year 2011 and continuing through the end of fiscal year 2021, each agency Plan shall:

(a) include a policy statement committing the agency to compliance with environmental and energy statutes, regulations, and Executive Orders;

(b) achieve the sustainability goals and targets, including greenhouse gas reduction targets, established under section 2 of this order;

(c) be integrated into the agency's strategic planning and budget process, including the agency's strategic plan under section 3 of the Government Performance and Results Act of 1993, as amended (5 U.S.C. 306);

(d) identify agency activities, policies, plans, procedures, and practices that are relevant to the agency's implementation of this order, and where necessary, provide for development and implementation of new or revised policies, plans, procedures, and practices;

(e) identify specific agency goals, a schedule, milestones, and approaches for achieving results, and quantifiable metrics for agency implementation of this order;

(f) take into consideration environmental measures as well as economic and social benefits and costs in evaluating projects and activities based on lifecycle return on investment;

(g) outline planned actions to provide information about agency progress and performance with respect to achieving the goals of this order on a publicly available Federal website;

(h) incorporate actions for achieving progress metrics identified by the OMB Director and the CEQ Chair;

(i) evaluate agency climate-change risks and vulnerabilities to manage the effects of climate change on the agency's operations and mission in both the short and long term; and

(j) identify in annual updates opportunities for improvement and evaluation of past performance in order to extend or expand projects that have net lifecycle benefits, and reassess or discontinue under-performing projects.

Sec. 9. Recommendations for Greenhouse Gas Accounting and Reporting. The Department of Energy, through its Federal Energy Management Program, and in coordination with the Environmental Protection Agency, the Department of Defense, the General Services Administration, the Department of the Interior, the Department of Commerce, and other agencies as appropriate, shall:

(a) within 180 days of the date of this order develop and provide to the CEQ Chair recommended Federal greenhouse gas reporting and accounting procedures for agencies to use in carrying out their obligations under subsections 2(a), (b), and (c) of this order, including procedures that will ensure that agencies:

more

(OVER)

10

(i) accurately and consistently quantify and account for greenhouse gas emissions from all scope 1, 2, and 3 sources, using accepted greenhouse gas accounting and reporting principles, and identify appropriate opportunities to revise the fiscal year 2008 baseline to address significant changes in factors affecting agency emissions such as reorganization and improvements in accuracy of data collection and estimation procedures or other major changes that would otherwise render the initial baseline information unsuitable;

(ii) consider past Federal agency efforts to reduce greenhouse gas emissions; and

(iii) consider and account for sequestration and emissions of greenhouse gases resulting from Federal land management practices;

(b) within 1 year of the date of this order, to ensure consistent and accurate reporting under this section, provide electronic accounting and reporting capability for the Federal greenhouse gas reporting procedures developed under subsection (a) of this section, and to the extent practicable, ensure compatibility between this capability and existing Federal agency reporting systems; and

(c) every 3 years from the date of the CEQ Chair's issuance of the initial version of the reporting guidance, and as otherwise necessary, develop and provide recommendations to the CEQ Chair for revised Federal greenhouse gas reporting procedures for agencies to use in implementing subsections 2(a), (b), and (c) of this order.

Sec. 10. Recommendations for Sustainable Locations for Federal Facilities. Within 180 days of the date of this order, the Department of Transportation, in accordance with its Sustainable Partnership Agreement with the Department of Housing and Urban Development and the Environmental Protection Agency, and in coordination with the General Services Administration, the Department of Homeland Security, the Department of Defense, and other agencies as appropriate, shall:

(a) review existing policies and practices associated with site selection for Federal facilities; and

(b) provide recommendations to the CEQ Chair regarding sustainable location strategies for consideration in Sustainability Plans. The recommendations shall be consistent with principles of sustainable development including prioritizing central business district and rural town center locations, prioritizing sites well served by transit, including site design elements that ensure safe and convenient pedestrian access, consideration of transit access and proximity to housing affordable to a wide range of Federal employees, adaptive reuse or renovation of buildings, avoidance of development of sensitive land resources, and evaluation of parking management strategies.

Sec. 11. Recommendations for Federal Local Transportation Logistics. Within 180 days of the date of this order, the General Services Administration, in coordination with the Department of Transportation, the Department of the Treasury, the Department of Energy, the Office of Personnel Management,

more

11

and other agencies as appropriate, shall review current policies and practices associated with use of public transportation by Federal personnel, Federal shuttle bus and vehicle transportation routes supported by multiple Federal agencies, and use of alternative fuel vehicles in Federal shuttle bus fleets, and shall provide recommendations to the CEQ Chair on how these policies and practices could be revised to support the implementation of this order and the achievement of its policies and goals.

Sec. 12. Guidance for Federal Fleet Management. Within 180 days of the date of this order, the Department of Energy, in coordination with the General Services Administration, shall issue guidance on Federal fleet management that addresses the acquisition of alternative fuel vehicles and use of alternative fuels; the use of biodiesel blends in diesel vehicles; the acquisition of electric vehicles for appropriate functions; improvement of fleet fuel economy; the optimizing of fleets to the agency mission; petroleum reduction strategies, such as the acquisition of low greenhouse gas emitting vehicles and the reduction of vehicle miles traveled; and the installation of renewable fuel pumps at Federal fleet fueling centers.

Sec. 13. Recommendations for Vendor and Contractor Emissions. Within 180 days of the date of this order, the General Services Administration, in coordination with the Department of Defense, the Environmental Protection Agency, and other agencies as appropriate, shall review and provide recommendations to the CEQ Chair and the Administrator of OMB's Office of Federal Procurement Policy regarding the feasibility of working with the Federal vendor and contractor community to provide information that will assist Federal agencies in tracking and reducing scope 3 greenhouse gas emissions related to the supply of products and services to the Government. These recommendations should consider the potential impacts on the procurement process, and the Federal vendor and contractor community including small businesses and other socioeconomic procurement programs. Recommendations should also explore the feasibility of:

(a) requiring vendors and contractors to register with a voluntary registry or organization for reporting greenhouse gas emissions;

(b) requiring contractors, as part of a new or revised registration under the Central Contractor Registration or other tracking system, to develop and make available its greenhouse gas inventory and description of efforts to mitigate greenhouse gas emissions;

(c) using Federal Government purchasing preferences or other incentives for products manufactured using processes that minimize greenhouse gas emissions; and

(d) other options for encouraging sustainable practices and reducing greenhouse gas emissions.

Sec. 14. Stormwater Guidance for Federal Facilities. Within 60 days of the date of this order, the Environmental Protection Agency, in coordination with other Federal agencies as appropriate, shall issue guidance on the implementation of section 438 of the Energy Independence and Security Act of 2007 (42 U.S.C. 17094).

more

(OVER)

12

Sec. 15. Regional Coordination. Within 180 days of the date of this order, the Federal Environmental Executive shall develop and implement a regional implementation plan to support the goals of this order taking into account energy and environmental priorities of particular regions of the United States.

Sec. 16. Agency Roles in Support of Federal Adaptation Strategy. In addition to other roles and responsibilities of agencies with respect to environmental leadership as specified in this order, the agencies shall participate actively in the interagency Climate Change Adaptation Task Force, which is already engaged in developing the domestic and international dimensions of a U.S. strategy for adaptation to climate change, and shall develop approaches through which the policies and practices of the agencies can be made compatible with and reinforce that strategy. Within 1 year of the date of this order the CEQ Chair shall provide to the President, following consultation with the agencies and the Climate Change Adaptation Task Force, as appropriate, a progress report on agency actions in support of the national adaptation strategy and recommendations for any further such measures as the CEQ Chair may deem necessary.

Sec. 17. Limitations. (a) This order shall apply to an agency with respect to the activities, personnel, resources, and facilities of the agency that are located within the United States. The head of an agency may provide that this order shall apply in whole or in part with respect to the activities, personnel, resources, and facilities of the agency that are not located within the United States, if the head of the agency determines that such application is in the interest of the United States.

(b) The head of an agency shall manage activities, personnel, resources, and facilities of the agency that are not located within the United States, and with respect to which the head of the agency has not made a determination under subsection (a) of this section, in a manner consistent with the policy set forth in section 1 of this order to the extent the head of the agency determines practicable.

Sec. 18. Exemption Authority.

(a) The Director of National Intelligence may exempt an intelligence activity of the United States, and related personnel, resources, and facilities, from the provisions of this order, other than this subsection and section 20, to the extent the Director determines necessary to protect intelligence sources and methods from unauthorized disclosure.

(b) The head of an agency may exempt law enforcement activities of that agency, and related personnel, resources, and facilities, from the provisions of this order, other than this subsection and section 20, to the extent the head of an agency determines necessary to protect undercover operations from unauthorized disclosure.

(c) (i) The head of an agency may exempt law enforcement, protective, emergency response, or military tactical vehicle fleets of that agency from the provisions of this order, other than this subsection and section 20.

more

13

(ii) Heads of agencies shall manage fleets to which paragraph (i) of this subsection refers in a manner consistent with the policy set forth in section 1 of this order to the extent they determine practicable.

(d) The head of an agency may exempt particular agency activities and facilities from the provisions of this order, other than this subsection and section 20, where it is in the interest of national security. If the head of an agency issues an exemption under this section, the agency must notify the CEQ Chair in writing within 30 days of issuance of the exemption under this subsection. To the maximum extent practicable, and without compromising national security, each agency shall strive to comply with the purposes, goals, and implementation steps in this order.

(e) The head of an agency may submit to the President, through the CEQ Chair, a request for an exemption of an agency activity, and related personnel, resources, and facilities, from this order.

Sec. 19. Definitions. As used in this order:

(a) "absolute greenhouse gas emissions" means total greenhouse gas emissions without normalization for activity levels and includes any allowable consideration of sequestration;

(b) "agency" means an executive agency as defined in section 105 of title 5, United States Code, excluding the Government Accountability Office;

(c) "alternative fuel vehicle" means vehicles defined by section 301 of the Energy Policy Act of 1992, as amended (42 U.S.C. 13211), and otherwise includes electric fueled vehicles, hybrid electric vehicles, plug-in hybrid electric vehicles, dedicated alternative fuel vehicles, dual fueled alternative fuel vehicles, qualified fuel cell motor vehicles, advanced lean burn technology motor vehicles, self-propelled vehicles such as bicycles and any other alternative fuel vehicles that are defined by statute;

(d) "construction and demolition materials and debris" means materials and debris generated during construction, renovation, demolition, or dismantling of all structures and buildings and associated infrastructure;

(e) "divert" and "diverting" means redirecting materials that might otherwise be placed in the waste stream to recycling or recovery, excluding diversion to waste-to-energy facilities;

(f) "energy intensity" means energy consumption per square foot of building space, including industrial or laboratory facilities;

(g) "environmental" means environmental aspects of internal agency operations and activities, including those aspects related to energy and transportation functions;

(h) "excluded vehicles and equipment" means any vehicle, vessel, aircraft, or non-road equipment owned or operated by an agency of the Federal Government that is used in:

more

(OVER)

14

(i) combat support, combat service support, tactical or relief operations, or training for such operations;

(ii) Federal law enforcement (including protective service and investigation);

(iii) emergency response (including fire and rescue); or

(iv) spaceflight vehicles (including associated ground-support equipment);

(i) "greenhouse gases" means carbon dioxide, methane, nitrous oxide, hydrofluorocarbons, perfluorocarbons, and sulfur hexafluoride;

(j) "renewable energy" means energy produced by solar, wind, biomass, landfill gas, ocean (including tidal, wave, current, and thermal), geothermal, municipal solid waste, or new hydroelectric generation capacity achieved from increased efficiency or additions of new capacity at an existing hydroelectric project;

(k) "scope 1, 2, and 3" mean;

(i) scope 1: direct greenhouse gas emissions from sources that are owned or controlled by the Federal agency;

(ii) scope 2: direct greenhouse gas emissions resulting from the generation of electricity, heat, or steam purchased by a Federal agency; and

(iii) scope 3: greenhouse gas emissions from sources not owned or directly controlled by a Federal agency but related to agency activities such as vendor supply chains, delivery services, and employee travel and commuting;

(l) "sustainability" and "sustainable" mean to create and maintain conditions, under which humans and nature can exist in productive harmony, that permit fulfilling the social, economic, and other requirements of present and future generations;

(m) "United States" means the fifty States, the District of Columbia, the Commonwealth of Puerto Rico, Guam, American Samoa, the United States Virgin Islands, and the Northern Mariana Islands, and associated territorial waters and airspace;

(n) "water consumption intensity" means water consumption per square foot of building space; and

(o) "zero-net-energy building" means a building that is designed, constructed, and operated to require a greatly reduced quantity of energy to operate, meet the balance of energy needs from sources of energy that do not produce greenhouse gases, and therefore result in no net emissions of greenhouse gases and be economically viable.

Sec. 20. General Provisions.

(a) This order shall be implemented in a manner consistent with applicable law and subject to the availability of appropriations.

more

15

(b) Nothing in this order shall be construed to impair or otherwise affect the functions of the OMB Director relating to budgetary, administrative, or legislative proposals.

(c) This order is intended only to improve the internal management of the Federal Government and is not intended to, and does not, create any right or benefit, substantive or procedural, enforceable at law or in equity by any party against the United States, its departments, agencies, or entities, its officers, employees, or agents, or any other person.

BARACK OBAMA

THE WHITE HOUSE,
October 5, 2009.

# # #

NOTES

FOREWORD

1. Paul W. Gates, *History of Public Land Law Development,* written for the Public Land Law Review Commission, Washington, DC, U.S. Government Printing Office, November 1968, p. 51, quoting Worthington C. Ford, *Journals of the Continental Congress,* vol. 18: 915.
2. Manuscript letters of Ferdinand Hayden to S. F. Baird, Smithsonian Institute Archives, RU 7002, February 16, 1853.
3. "Habitat," Section D., Land (Agenda Item 10 (d)), Preamble, 2. United Nations Conference on Human Settlements, Vancouver, May 31–June 11, 1976.
4. Federal Land Planning and Management Act, Section 205.

CHAPTER 1

1. Friedrich Engels, *Ludwig Feuerbach,* 1886.
2. Carl Sagan, *Cosmos,* TV mini-series documentary, 1980.
3. Friedrich Engels, *Anti-Duhring,* 1878.
4. "The All-new 2010 Toyota Prius: Ready to Welcome More Canadians to the Sustainable Transportation Revolution," press release, Toyota Canada, Inc., January 12, 2010.
5. "Certified Sustainable Seafood," Whole Foods Market Values and Action, retrieved from Whole Foods website, March 6, 2011, http://www.wholefoodsmarket.com/values/certified-sustainable.php.
6. Engels, *Anti-Duhring.*
7. George Washington, Farewell Address, September 17, 1796.
8. John Adams, letter to Mercy Warren, April 16, 1776.
9. Patrick Henry, speech to the legislative body of the Virginia Colony, March 13, 1775.
10. C. H. Millar, *Florida, South Carolina, and Canadian Phosphates* (London: Eden Fisher and Co., 1892), 15.
11. The translation of this passage from the introduction to the 1862 edition of von Liebig's book

follows Erland Marold in "Everything Circulates: Agriculture, Chemistry, and Recycling Theories in the Second Half of the Nineteenth Century," *Environment and History* 8 (2002): 74.

12. Letter from Marx to Engels, February 13, 1866, from Marx and Engels, *Selected Correspondence* (New York: International Publishers, 1942), 204–5.
13. Karl Marx, *Das Kapital*, vol. 1, pt. 5, chap. 16, "Absolute and Relative Surplus Value."
14. Ibid., vol. 3, pt. 1, chap. 6, "The Effect of Price Fluctuation," sect. 2, http://www.marxists.org/archive/marx/works/1894-c3/ch06.htm.
15. Ibid., vol. 1, chap. 24, "Conversion of Surplus-Value into Capital," sect. 4, http://www.marxists.org/archive/marx/works/1867-c1/ch24.htm.
16. Ibid., vol. 3, pt. 1, chap. 6, "The Effect of Price Fluctuation," sect. 2, http://www.marxists.org/archive/marx/works/1894-c3/ch06.htm.
17. Ibid., vol. 1, chap. 15, "Machinery and Modern Industry," sect. 10, http://www.marxists.org/archive/marx/works/1867-c1/ch15.htm#a245.
18. John Bellamy Foster, "Marx Ecology in Historical Perspective," *International Socialist Journal*, Issue 96 (Winter 2002).
19. Ibid.
20. Edwin Ray Lankester, *Nature and Man* (Oxford Clarendon Press, 1905), 23.
21. Ibid., 27.
22. Arthur G. Tansley, "The Use and Abuse of Vegetational Concepts and Terms," *Ecology* 16, no. 3 (July 1935): 299, 303–4.
23. Charles Elton, *The Ecology of Invasions by Animals and Plants* (London: Methuen and Co., 1958), 137–42.
24. The text for the Decree on Land and the Decree on Forests can be accessed on *Wikipedia*, http://en.wikipedia.org/wiki/Soviet_Decree.
25. For more on the decree "On Hunting Seasons and the Right to Possess Hunting Weapons," see Louis Proyect, "The Ecology of Nazis, Greens and Socialists," *Socialist Viewpoint* vol. 4, no. 3, March 2004, http://www.socialistviewpoint.org/mar_04/mar_04_19.html.
26. Douglas Weiner, *Models of Nature: Ecology, Conservation and Cultural Revolution in Soviet Russia* (Pittsburgh: University of Pittsburgh, 2000), 27.
27. R. J. Overy, *The Dictators: Hitler's Germany and Stalin's Russia*. W. W. Norton & Company, Inc., 2004), 399.
28. Adolf Hitler, *Mein Kampf*, chap. 10.
29. Adolf Hitler, quoted in Raymond H. Dominick III, *The Environmental Movement in Germany: Prophets and Pioneers, 1871–1971* (Bloomington, IN: Indiana University Press, 1992), 81.

CHAPTER 2

1. Van Jones, keynote address, Power Shift Conference on climate change, Washington, D.C., February 27, 2009.
2. This pledge was famously taken by members of the National Student League during the student demonstrations against American involvement in World War I on April 13, 1934.
3. Only one edition of *Spark* was published; I have a copy, dated June, 1932.
4. A. G. Tansley, "The Use and Abuse of Vegetational Concepts and Terms," *Ecology* 16, no. 3 (July 1935): 303–4.
5. Charles Elton, *The Ecology of Invasions by Animals and Plants* (London: Methuen and Co., 1958), 137–42.
6. Rachel Carson, *Silent Spring* (n.p.: Houghton Mifflin Co., 1962), 155.
7. John Bellamy Foster, "Rachel Carson's Ecological Critique," *Monthly Review*, January 2008.
8. John Bellamy Foster and Brett Clark, "Rachel Carson's Ecological Critique," *Monthly Review*,

February, 2008.
9. Rachel Carson, *Lost Woods* (1964; repr., Boston: Beacon, 1998), 194.
10. United States Forest Service, "Rachel Carson's National Wildlife Refuge," http://www.fws.gov/northeast/rachelcarson/carsonbio.html; http://www.fws.gov/rachelcarson/RC_Conservation_Legacy.pdf.
11. J. Gordon Edwards, "Mosquitoes, DDT, and Human Health, *21st Century Magazine*, Fall 2002, http://www.21stcenturysciencetech.com/articles/Fall02/Mosquitoes.html.
12. "Rachel Carson: A Conservation Legacy," United States Fish and Wildlife Service, Office of External Affairs, http://www.fws.gov/rachelcarson/.
13. Paul Ehrlich, *The Population Bomb* (New York: Ballantine Books, 1968), 1.
14. Ibid.
15. Ibid., 66.
16. Ibid., xi–xii.
17. *Democracy Now* radio broadcast from Stanford University, September 6, 2007.
18. H. Lanier Hickman Jr., *American Alchemy: A History of Solid Waste Management in the United States* (n.p.: Forester Press, 2003), 520.
19. "History of Earth Day," Community Environmental Council, Santa Barbara, http://www.communityenvironmentalcouncil.org/Events/Earthday/. No longer accessible.
20. History 179, North American Environmental History, Lecture 20, Brown University, http://www.brown.edu/Courses/HI0179/Lectures/Lecture_Twentyone.htm.
21. Richard Seven, "Treading Lightly One Small Step at a Time," *Seattle Times*, April 21, 2002.
22. Ibid.
23. Gaylord Nelson, "How the First Earth Day Came About," *American Heritage Magazine*, October, 1993.
24. Tim Brown, "Earth Day and the Rise of Environmental Consciousness," United States International Information Programs, State Department, April 11, 2005, http://usinfo.state.gov/gi/Archive/2005/Apr/11-390328.html. No longer accessible.
25. "IMPLEMENTATION OF AGENDA 21: REVIEW OF PROGRESS MADE SINCE THE UNITED NATIONS CONFERENCE ON ENVIRONMENT AND DEVELOPMENT, 1992," prepared by the Government of the United States for presentation to the United Nations Commission on Sustainable Development, April 7–25, 1997, New York City, http://www.un.org/esa/earthsummit/usa-cp.htm.
26. Reorganization Plan Number 3, July 9, 1970, http://www.epa.gov/history/org/origins/reorg.htm. No longer accessible.
27. This PSA can be viewed on YouTube at http://www.youtube.com/watch?v=lR06-RP3n0Q.
28. Snopes, "Iron Eyes Cody," http://www.youtube.com/watch?v=lR06-RP3n0Q.
29. Frederick Engels, *Dialectics of Nature* (1883), introduction, http://www.marxists.org/archive/marx/works/1883/don/ch01.htm.
30. Ehrlich, *Population Bomb*, 51.
31. Between 1940 and 1970 the average surface temperature of the planet decreased by .18 degrees Fahrenheit (.1 degree Celsius).
32. Stephen Schneider, *Discover*, October 1989, 44–45.
33. MauriceStrong.net, "Short Biography," http://www.mauricestrong.net/2008072115/strong-biography.html.
34. Ibid.
35. Maurice Strong, 1972 interview on BBC, http://www.youtube.com/watch?v=1YCatox0Lxo&feature=player_embedded.
36. Ronald Bailey, "Who Is Maurice Strong?" *National Review*, September 1, 1997.
37. Ezra Levant, "Kyoto Protocol Compiled by Un-elected Global Bureaucrats," *Calgary Sun*, December 2, 2002.

38. "Habitat," Section D., Land (Agenda Item 10 (d)), Preamble, 1, United Nations Conference on Human Settlements, Vancouver, May 31–June 11, 1976.
39. Ibid. Preamble, 2.
40. Ibid. Preamble, 3 (b) (i).
41. Ibid. Preamble, 3 (b) (iv).
42. John Holdren, Anne Ehrlich, and Paul Ehrlich, *Ecoscience, Population, Resources, Environment* (San Francisco: W.H. Freeman and Company, 1977), 837.

CHAPTER 3

1. Georg Hegel, *Philosophy of Right,* part 3: Ethical Life, The State, section 258 (1820).
2. Karl Marx, *Eleventh Thesis on Feuerbach* (1845), emphasis added.
3. "Neutralist with Moral Fiber," *Time,* November 10, 1961.
4. U Thant, "Ten Crucial Years" (address to the United Nations General Assembly), May 9, 1969.
5. Jack Lewis, "The Birth of the EPA," *EPA Journal,* November, 1985.
6. Gandhi certainly believed in communism and socialism and once wrote in the English newspaper *Amrita Bazar Patrika* on August 3, 1934, "Our socialism or communism should, therefore, be based on nonviolence and on harmonious co-operation of labour and capital, landlord and tenant."
7. "1972 United Nations Conference on the Human Environment (Part 1)," http://www.youtube.com/watch?v=mJUk70tfELA&feature=player_embedded#!.
8. Declaration of the United Nations Conference on the Human Environment, June 16, 1972.
9. From an obscure news film documenting the UN Conference on Human Environment and presented on YouTube: http://www.youtube.com/watch?v=mJUk70tfELA.
10. President Richard Nixon, Statement 206, "Statement about the United Nations Environment Conference on the Human Environment in Stockholm, Sweden, June 20, 1972.
11. Ibid.
12. President Richard Nixon, Statement 360, "Statement about Signing the United Nations Environment Program Participation Act of 1973," December 17, 1973.
13. "World Charter for Nature," United Nations, 48th plenary meeting, October 28, 1982.
14. Ibid.
15. Office of Congresswoman Nancy Pelosi, "BART to the Airport Funding," October 9, 1997, http://pelosi.house.gov/pressarchives/releases/prtransp.htm.
16. Michael Cabanatuan, "Squashed Endangered Garter Snake Quashes Progress on Extension," *San Francisco Chronicle,* May 11, 2002.
17. United Nations Resolution 38/61, December 19, 1983.
18. H. George Fredrickson, "Public Administration and Social Equity," *Public Administration Review* 50, no. 2 (March–April, 1990), 228–37.
19. Ibid.
20. *Our Common Future: Report of the World Commission on Environment and Development,* United Nations, March 20, 1987, chap. 2, IV.1.
21. Ibid., IV.3.
22. Ibid., Overview, I.3.29.
23. Catherine Johnson, "Our Common Future," *Canadian Business Review,* March 22, 1990.
24. United Nations Environmental Program, Agenda 21, chap. 6.13.
25. Ibid., chap. 9.9.
26. William K. Reilly, "The Road from Rio," *EPA Journal,* September/October 1992, posted on the Environmental Protection Agency website, http://www.epa.gov/aboutepa/history/topics/summit/01.html.

27. Agenda 21, chap. 4, 4.11.
28. Ibid., 4.15.
29. Ibid., 4.23.
30. Lisa Jackson, on *Real Time with Bill Maher*, HBO, October 4, 2009.
31. Agenda 21, chap. 5, 5.3.
32. Ibid., 5.51.
33. Secretary of State Hillary Rodham Clinton, "Remarks on the 15th Anniversary of the International Conference on Population and Development," January 8, 2010.
34. Agenda 21, chap. 7, 7.20.
35. "Obama to Spend Billions for Green Jobs, Training," Associated Press, May 26, 2009.
36. Ibid.
37. America Clean Energy and Security Act of 2009, June 22, 2009, 323.
38. Ibid., 319.
39. Agenda 21, Chapter 7, 7.51.
40. Van Jones, radio interview with Amy Goodman, *Democracy Now!* October 28, 2008.
41. Van Jones, Uprising Radio, April, 2008, http://www.youtube.com/watch?v=3wh6AWCJdxg.
42. Kevin Freking, "Amtrak Loss Comes to $32 Per Passenger: Study," *Huffington Post,* October 27, 2009.
43. Agenda 21, Chapter 7, 7.52.
44. Ibid., 7.69.
45. http://www.recovery.gov/espsearch/Pages/default.aspx?k=pedestrian+path.
46. Agenda 21, Chapter 7, 7.52.
47. Ibid., 7.69.

CHAPTER 4

1. Micah Morrison, "Al Gore Jr.: Occidental Oriental Connection," *Wall Street Journal*, June 29, 2000.
2. Timothy Noah, "Was Albert Gore Sr. a Crook?" *Slate,* April 6, 2000, http://www.slate.com/id/78634/.
3. Morrison, "Al Gore Jr."
4. Micah Morrison, "Al Gore, Environmentalist and Zinc Miner," *Wall Street Journal*, June 29, 2000.
5. Neil Lyndon, "How Mr. Clean Got His Hands Dirty," *Sunday London Telegraph*, November 1, 1998.
6. George Starke, "Nagasaki after August 9, 1945," *St. Helena Star*, August, 9, 2000. George Starke is noted on the International Physicians for the Prevention of Nuclear War website as a "long-time supporter" of their cause, http://www.ippnw.org/News/Articles/2009StarkeAug.htm. No longer accessible.
7. Agenda 21, chap. 8, 8.21.
8. President William Jefferson Clinton, "Remarks in a Teleconference on Empowerment Zones and an Exchange with Reporters – Transcript," Weekly Compilation of Presidential Documents, May 10, 1993.
9. http://www.wri.org/about/strategic-plan.
10. Natural Resources Defense Council, Mission Statement.
11. Sierra Club Initiative to Limit Total Greenhouse Emissions, http://www.sierraclub.org/carbon/.
12. Nature Conservancy, "Climate Change: What We Do," http://www.nature.org/initiatives/climatechange/strategies/art13748.html.
13. World Wildlife Fund, "Who We Are: Vision," http://www.worldwildlife.org/who/Vision/

index.html.

14. Joseph Spector, *Politics on the Hudson*, "Cuomo Leads 12 States in Lawsuit Against EPA Over Global Warming," August 25, 2008, http://polhudson.lohudblogs.com/2008/08/25/cuomo-leads-12-states-in-lawsuit-against-epa-over-global-warming/.
15. Brad Knickerbocker, "Where the '08 Contenders Stand on Global Warming," *Christian Science Monitor,* October 15, 2007.
16. Dan Morgan, "Enron Also Courted Democrats," *Washington Post,* January, 13, 2002, A1.
17. Ibid.
18. President's Council on Sustainable Development, *Sustainable America—A New Consensus,* We Believe Statement, number 10, The White House, 1995.
19. Ibid., Number 11.
20. Ibid., Number 12.
21. President's Council on Sustainable Development, Revised Charter, Scope of Activities, April 25, 1997.
22. Al Gore, blog entry, "Fox News Manipulates Climate Coverage," December 15, 2010.
23. Al Gore, interview with Leslie Stahl, *60 Minutes,* CBS Television Network, March 30, 2008.
24. President's Council on Sustainable Development, *Sustainable America—A New Consensus*, chap. 5, Policy Recommendation 1, Action 4, February 1996.
25. "On Being a Scientist: Responsible Conduct in Research," National Academy of Science, 1996.
26. Agenda 21, chap. 8, 8.4.
27. Vice President Al Gore, letter to participants attending the National Environmental Monitoring and Research Workshop, September 25, 1996, accessed from the Environmental Division of the White House Office of Science and Technology Policy, http://dieoff.org/page51.htm.
28. Hearing on H.R. 1845, the National Biological Survey Act of 1993, September 14, 1993.
29. "Modern State Planning Statutes, The Growing Smart Working Papers," vol. 1, American Planning Association, Planning Advisory Service, Report Number 462/463, March 1996.
30. Agenda 21, Chapter 8, 8.13.
31. Ibid., Agenda 21, Chapter 8, 8.13.
32. "Implementation of Agenda 21: Review of the Progress Made Since the United Nations Conference on Environment and Development, 1992." prepared by the Government of the United States for presentation to the United Nations Commission on Sustainable Development, April 7–25, 1997, New York City, http://www.un.org/esa/earthsummit/usa-cp.htm.
33. Joel Schwartz, "Facts Not Fear on Air Pollution," NCPA Policy Report #294, December 2006, ISBN #1-56808-167-7.
34. John Christie, "My Nobel Moment," *Wall Street Journal*, November 1, 2007.
35. NBC Television, *Today Show,* November 5, 2007.
36. "Ex-Astronaut: Global Warming Is Bunk," Associated Press, February 16, 2009.
37. U.S. Senate Minority Report, Senator James Inhofe, Senate Committee on Environment and Public Works, March 17, 2009.
38. Harold Lewis, resignation letter to Curtis Callan, president of the American Physical Society, posted online by James Delingpole, blogger, *London Telegraph*, October 9, 2011, http://blogs.telegraph.co.uk/news/jamesdelingpole/100058265/us-physics-professor-global-warming-is-the-greatest-and-most-successful-pseudoscientific-fraud-i-have-seen-in-my-long-life/.
39. Al Gore, Aspen Institute Media Forum, Aspen, Colorado, August 4, 2011, audio provided by the *Colorado Independent*'s website, August 9, 2011, http://coloradoindependent.com/95685/audio-al-gore-calling-out-dissenters-on-climate-change.
40. Jeffrey H. Birnbaum, "Al Gore's Clinton Moment: His Protests of Innocence About the Buddhist Temple Fundraiser Look Very Squishy," *Fortune* magazine, February 7, 2000.
41. Stockpickr looked at twenty-nine known positions held by Generation Investment Management, as of the fourth quarter of 2010, http://www.stockpickr.com/pro/portfolio/al-gore-generation-

investment-management/. My original valuation of GIM was $1 Billion and is found in Brian Sussman, *Climategate* (Washington DC: WND Books, 2010), 206.
42. Alexander Haislip and Dan Primack, "Kleiner Perkins Raising Green Growth Fund," *Private Equity Week*, April 24, 2008, http://www.pewnews.com/story.asp?sectioncode=36&storycode=44384.
43. John. M. Broder, "Gore's Duel Role: Advocate and Investor," *New York Times*, November 2, 2009.
44. "Kleiner Perkins Caufield & Beyers Lead $75 Million Investment in Silver Springs Networks, Smart Grid Technology Leader," *Business Wire*, October 7, 2008, http://findarticles.com/p/articles/mi_m0EIN/is_2008_Oct_7/ai_n29480769/.
45. Miles Weiss, "Gore Invests $35 Million for Hedge Fund with EBAY Billionaire," *Bloomberg*, March 6, 2008, http://www.bloomberg.com/apps/news?pid=20601070&sid=a7li9Nhmhvg0&refer=home.
46. Broder, "Gore's Duel Role: Advocate and Investor."

CHAPTER 5

1. Mann apparently inputted data collected by Dr. Shaopeng Huang of the University of Michigan, who examined six thousand tree-ring boreholes from around the world. Like Dr. Deming's research, Huang's indicated that a pronounced global warming was evident in the medieval period, with temperatures significantly warmer than today. See: Huang, Pollack and Shen, "Late Quaternary Temperature Changes Seen in Worldwide Continental Heat Flow Measurements," *Geophysical Research Letters* 24: 1947–1950, 1997.
2. M. E. Mann, R. S. Bradley, and M. K. Hughes, "Global-Scale Temperature Patterns and Climate Forcing over the Past Six Centuries," *Nature* 392 (1998): 779–87.
3. M. E. Mann, et al., "Northern Hemisphere Temperatures during the Past Millennium: Inferences, Uncertainties, and Limitations," AGU *Geophysical Research Letters* (3.1) 26, no. 6 (1999): 759–62.
4. IPCC Technical Summary, 2001, 29.
5. Jonathan Petre, "Climategate U-Turn: Astonishment as Scientist at Centre of Global Warming Email Now Admits There Has Been no Global Warming Since 1995," *London Daily Mail*, February 14, 2010.
6. Joseph D'Aleo, Anthony Watts, "Surface Temperature Records: Policy Driven Deception?" Science and Public Policy Institute, Original Paper, originally published June 2, 2010.
7. Includes cars, trucks, and buses, but does not include off-road vehicles (Wards Auto, August 15, 2011, http://wardsauto.com/ar/world_vehicle_population_110815/).
8. International Panel on Climate Change, Fourth Assessment Report, 2007, chap. 5, 409.
9. Ibid., chap. 10, 821.
10. Abdulla Naseer, "Status of Coral Mining in the Maldives: Impacts and Management Options," Marine Research Section, Ministry of Fisheries and Agriculture Malé, Republic of Maldives (paper located on Food and Agriculture Organization of the United Nations website), June 16, 2009, http://www.fao.org/docrep/X5623E/x5623e0o.htm.
11. Ibid.
12. Ibid.
13. Mark Chipperfield in Tuvalu and David Harrison in London, "Falling Sea Level Upsets Theory of Global Warming," Telegraph (London), August 6, 2000.
14. Nils-Axel Mörner, Michael Tooley, and Göran Possnert, "New Perspectives for the Future of the Maldives," Global and Planetary Change 40, issues 1–2 (January 2004): 177–82.
15. James E. Hester, personal email, December 2000, from John Daly, "Top of the World: Is the North Pole Turning to Water?" http://www.john-daly.com/polar/arctic.htm.

16. M. Ogi, K. Yamazaki, and J. M. Wallace, "Influence of Winter and Summer Surface Wind Anomalies on Summer Arctic Sea Ice Extent," *Geophysical Research. Letters* 37, L07701, 2010.
17. Gadiosa Lamtey, "Mt. Kilimanjaro ice still baffles experts," *Guardian*, November 26, 2007.
18. Ibid.
19. Sussman, *Climategate*, 387.
20. C. J. Yapp, and H. Poths, "Ancient Atmospheric CO_2 Pressures Inferred from Natural Geothites," *Nature* 355, no. 23 (January 1992), 342–44.
21. Sylan H. Wittwer, "Rising Carbon Dioxide Is Great for Plants," *Policy Review*, Fall 1992.
22. Ibid.
23. Fischer et al.,"Ice Core Records of Atmospheric CO_2 Around the Last Three Glacial Terminations," *Science*, March 12, 1999.
24. "Hansen Calls for Key to Lead on Climate Change," *Standard*, May 25, 2011, http://thestandard.org.nz/hansen-calls-for-key-to-lead-on-climate-change/.
25. Senate Bill 2433, "Global Poverty Act," Senator Barack Obama, lead sponsor, introduced December 7, 2007.
26. Larry Petrash, "Reject Obama's Senate Bill S. 2433," *Times Record News*, August 3, 2008.
27. Vincent Gioia, "United Nations' Power Will Grow under Obama," *Post Chronicle*, December 31, 2008.
28. Alister Doyle, "$750 Billion 'Green' Investment Could Revive Economy: U.N." Reuters, March 19, 2009.
29. Deborah Corey Barnes, "The Money and Connections behind Al Gore's Carbon Crusade," *Human Events*, October 3, 2000, http://www.humanevents.com/article.php?id=22663.
30. http://www2.goldmansachs.com/citizenship/environment/business-initiatives.html, as of August 30, 2011.
31. Karan Cappor and Phillippe Ambrosi, "State and Trends of the Carbon Market 2008," World Bank, May 2008.
32. Michael Szabo, "ICE Cuts Staff at Chicago Climate Exchange: Sources," Reuters, August 11, 2010, http://www.reuters.com/article/2010/08/11/us-carbon-ccx-layoffs-idUSTRE67A3K620100811.
33. Ibid.
34. Ibid.
35. Addendum to Energy Policy Act of 2005, H.R. 6–370, sect. 1252, "Smart Metering."
36. Ibid.
37. Barack Obama's campaign speech, Lansing, MI, August 4, 2008.
38. Energy Policy Act of 2005, Subtitle B—Transmission Infrastructure Modernization, Section 1221, Siting of Interstate Electric Transmission Facilities, amended to include Section 216, p. 354.
39. "DeWeese criticizes DOE Decision on Power Line," press release, Pennsylvania Majority Leader Bill DeWeese, October 2, 2007.
40. "DeWeese Seeks Repeal of Federal Energy Act," press release, Pennsylvania Majority Leader Bill DeWeese, May 11, 2007.
41. Amy Abel, Specialist in Energy Policy, "Smart Grid Provisions in H.R. 6, 100th Congress," Congressional Research Service, December 20, 2007.

CHAPTER 6

1. Robert Lanza, "Biocentrism: How Life Created the Universe," MSNBC.com, June 6, 2009.
2. Robert Lanza, "A New Theory of the Universe," *American Scholar*, http://www.theamericanscholar.org/a-new-theory-of-the-universe/.

3. Lanza, "Biocentrism."
4. "Environmental Ethics," *Stanford University Encyclopedia of Philosophy*, first published June 3, 2002; substantive revision January 3, 2008, http://plato.stanford.edu/entries/ethics-environmental/.
5. Ibid.
6. Global Forum of Spiritual and Parliamentary Leaders on Human Survival, *Shared Vision* 3, no. 1 (1989): 3.
7. Ibid.
8. United Nations Environment Program, *Sabbath Newsletter* 1, no. 2, Winter 1990, 1.
9. *Shared Vision* magazine 4, 1990, 4.
10. Green Cross International, Mission Statement, http://www.greencrossinternational.net/en/who-we-are/mission.
11. Mikhail Gorbachev, transcribed from the *Charlie Rose* television program, PBS, October 23, 1996.
12. National Religious Partnership for the Environment, "The Founding of the Partnership: 1990–1993," http://www.nrpe.org/whatisthepartnership/founding_intro03.htm#top. No longer accessible.
13. Henry Lamb, "The Rise of the Global Green Religion," Environmental Conservation Organization, 1996.
14. Ibid.
15. Ibid.
16. National Religious Partnership for the Environment, "Time Line," http://www.nrpe.org/whatisthepartnership/timeline03.htm#top. No longer accessible.
17. http://www.wheaton.edu/envstudies/es%20brochure.pdf. No longer accessible.
18. Al Gore, *Earth in the Balance* (New York: Houghton Mifflin, 1992), 269, 274.
19. Secretary of Education Arne Duncan, address, "Sustainability Education Summit: Citizenship and Pathways for a Green Economy," Washington Plaza Hotel Washington, D.C., September 21, 2010.
20. http://www.youthcannetwork.org/Youth_CAN_Network/Network_Initiatives.html.
21. June Q. Wu, "Student Ideas for Green Roof Make School a Teaching Lab," *Boston Globe*, June 26, 2010.
22. Boston Latin, Academic Programs, Course Description, http://www.bls.org/podium/default.aspx?t=113964&did=15806,D.
23. Children's Environmental Literacy Foundation, "Sustainable Sustainability Programs," http://celfeducation.org/About.html.
24. Leslie Kaufman, "A Cautionary Video about America's 'Stuff'," *New York Times*, May 10, 2009.
25. "Let There Be... Stuff? A Spirit-Filled Response to a Consumer-Crazed World: A Faith-Based Program for Christian Teens," Session 5, Waste Not, produced by The Story of Stuff Project and Green Faith.
26. Representative Debbie Wasserman-Shultz, speaking on CNN, January 10, 2011.
27. Al Gore, *An Inconvenient Truth: The Planetary Emergency of Global Warming and What We Can Do about It* (n.p.: Rodale Books, 2006).
28. James Hansen, "Coal-fired Power Stations are Death Factories. Close Them," *Observer*, February 15, 2009.
29. Interview with Maria Gilardin, TUC Radio, San Francisco, December 2006.
30. Dr. Willie Soon, "Endangering the Polar Bear,' speech presented at the Doctors for Disaster Preparedness annual conference, Mesa, Arizona, 2008.
31. "Frequently Asked Questions about Global and Climate Change. Back to Basics," United States Environmental Protection Agency, Office of Air and Radiation, EPA-430-ROB-016, April, 2009.

32. "Climate Change and Health Effects," United States Environmental Protection Agency, http://www.epa.gov/climatechange/downloads/Climate_Change_Health.pdf.
33. Aaron Klein, "Czar: 'Spread the wealth! Change the Whole System,'" World Net Daily, August 30, 2009, http://www.wnd.com/?pageId=108441#ixzz1KrkOduXB.
34. Sarah-Kate Templeton, "Two Children Should Be Limit, Says Green Guru," *London Sunday Times*, February 1, 2009.
35. See "Earth Day Idiots Exposed" on YouTube, http://www.youtube.com/watch?v=hRdJGQp6kjE.

CHAPTER 7

1. The White House, Office of the Press Secretary, "Remarks by the President on the Economy," Solyndra, Inc., Fremont, CA, May 26, 2010, http://www.whitehouse.gov/the-press-office/remarks-president-economy-0.
2. ABC News, "Connected Energy Firm Got Lowest Interest Rate on Government Loan," September 7, 2001, http://abcnews.go.com/Blotter/solyndra-lowest-interest-rate/story?id=14460246.
3. Vice President Joe Biden, address at Solyndra Headquarters, Fremont, California, September 4, 2009.
4. Secretary of Energy Steven Chu, address at Solyndra Headquarters, Fremont, California, September 4, 2009.
5. "Rudolph and Sletten to Build New Solar Manufacturing Facility for Solyndra in California," *Energy Business Review*, September 21, 2009, http://solar.energy-business-review.com/news/rudolph_and_sletten_to_build_new_solar_manufacturing_facility_for_solyndra_in_california_090921.
6. The White House, "Remarks by the President on the Economy."
7. Yuliya Chermova, "Solyndra CEO: We Made Two Major Mistakes," *Wall Street Journal*, March 3, 2011.
8. Congressman Pete Stark, House of Representatives, October 18, 2007.
9. Rudolph and Sletten Inc. accepted a $733 million contract in September 2009 to construct Solyndra's 603,000-square-foot Fab 2 facility. Simple math would place the per-square foot price at $1204 ("Rudolph and Sletten to Build New Solar Manufacturing Facility for Solyndra in California).
10. Amanda Carey, "Solyndra Officials Made Numerous Trips to the White House, Logs Show," *Daily Caller*, September 8, 2011, http://dailycaller.com/2011/09/08/solyndra-officials-made-numerous-trips-to-the-white-house-logs-show/.
11. William McQuillen, "Taxpayers Stand behind Solyndra Investors under Obama Deal, *Businessweek,* September 3, 2011, http://www.businessweek.com/news/2011-09-03/taxpayers-rank-behind-solyndra-investors-under-obama-deal.html.
12. Carey, "Solyndra Officials Made Numerous Trips to the White House, Logs Show."
13. Ronnie Green and Matthew Mosk, "Obama Fundraiser Got Clean Energy Aid, then perch to advise Energy Secretary," Center for Public Integrity, March 30, 2011, http://www.iwatchnews.org/2011/03/30/3845/green-bundler-golden-touch.
14. Westly Group, "What We Do," http://westlygroup.com/what-we-do/.
15. "Solar Company Failure Shows Less Federal Aid Works Better," Editorial, Bloomberg news Service, September 8, 2011, http://www.bloomberg.com/news/2011-09-08/solar-company-failure-shows-that-less-federal-aid-can-work-better-view.html.
16. Jessica Lillian, "Massachusetts Outrage: Evergreen Solar's Exodus to China," *Solar Industry Magazine,* January 18, 2011, http://solarindustrymag.com/e107_plugins/content/content.php?content.7085.
17. Evergreen Solar Press Release, "EVERGREEN SOLAR TO DISPLAY "MADE IN USA"

ES-A SERIES STRING RIBBON® SOLAR PANELS AT SOLAR POWER INTERNATIONAL IN LOS ANGELES, CALIF., OCTOBER 12–14, 2010," October 12, 2010, http://evergreensolar.com/en/2010/10/evergreen-solar-to-display-"made-in-usa"-es-a-series-string-ribbon®-solar-panels-at-solar-power-international-in-los-angeles-calif-october-12-14-2010/.
18. "Solar Company That Obama Visited Will Shut Down," Reuters , Wednesday August 31, 2011, http://www.reuters.com/article/2011/08/31/us-solyndra-idUSTRE77U5K420110831.
19. Energy Information Administration, "Monthly Energy Review," Table 9.9 Average Retail Prices of Electricity, August 29, 2011.
20. Nuclear Energy Institute, "U.S. Electricity Production Costs (1995-2010)."
21. Patrick McGreevy, "California Senate OKs Renewable Energy Bill," *Los Angeles Times,* February 25, 2011.
22. "More N.J. Homeowners Invest in Rooftop Solar Systems with Government Incentives, Cheaper Costs," Associated Press, October 21, 2009, http://www.nj.com/news/index.ssf/2009/10/more_nj_homeowners_invest_in_r.html.
23. Ayesha Rascoe, "BrightSource Solar Gets $1.6 Bln Loan Guarantee," Reuters , April 11, 2011, http://www.reuters.com/article/2011/04/11/brightsource-loan-guarantee-idUSN1125898820110411.
24. Josh Mitchell and Stephen Power, "Gore-Backed Car Firm Gets Large U.S. Loan," *Wall Street Journal,* September 25, 2009, http://online.wsj.com/article/SB125383160812639013.html#printMode.
25. Ibid.
26. Energy Information Administration, Department of Energy, "Federal Financial Interventions and Subsidies in Energy Markets 2007," http://www.eia.doe.gov/oiaf/servicerpt/subsidy2/pdf/execsum.pdf, accessed September 15, 2011.
27. "More N.J. Homeowners Invest in Rooftop Solar Systems with Government Incentives, Cheaper Costs."
28. This figure provided by solarpowerrocks.com.
29. United States Department of Energy, Sunshot Initiative, "DOE Announces $27 Million to Reduce Costs of Solar Energy Projects, Streamline Permitting and Installations," June 1, 2011, http://www1.eere.energy.gov/solar/sunshot/news_detail.html?news_id=17408.
30. Energy Information Administration, "2010 International Energy Outlook-Electricity," Report #: DOE/EIA-0383 (2010) May 11, 2010, http://www.eia.gov/oiaf/ieo/electricity.html.
31. Sean Corcoran, "The Falmouth Experience, Life under the Blades," WGBH, March 7, 2011.
32. Ibid.
33. Aaron Gouvela, "Turbine Shutdown Could Cost Town $173,000," *Cape Cod Times,* March 8, 2011.
34. "Bird Fatality Study at Altamont Pass Wind Resource Area, October 2005– September 2007," Altamont Pass Avian Monitoring Team, Draft Report, January 25, 2008.
35. "Wind Now Employs More People Than Coal," *Huffington Post,* January 29, 2009, http://www.huffingtonpost.com/2009/01/29/wind-now-employs-more-peo_n_162277.html.
36. "Coal and Jobs in the United States," Source Watch, http://www.sourcewatch.org/index.php?title=Coal_and_jobs_in_the_United_States.
37. Al Gore, remarks presented at the opening plenary session of the Clinton Global Initiative Annual Meeting, New York City, September 23, 2008.
38. Professor Michael J. Trebilcock, "Wind Power Is a Complete Disaster," comments made before the Ontario legislature on April 7, 2009, and published as an editorial in the *Financial Post,* April 8, 2009.

CHAPTER 8

1. U.S. Department of Interior, "Salazar Meets with BP Officials and Engineers at Houston Command Center to Review Response Efforts, Activities," press release, May, 6, 2010.
2. Noelle Straub, "Interior Approves Drilling on 2 Beaufort Seas Leases," *E&E News*, October 19, 2009, http://www.eenews.net/public/eenewspm/2009/10/19/3.
3. "Goodlatte Bill Would Push Drilling off Virginia," *Augusta Free Press*, April 6, 2011, http://augustafreepress.com/2011/04/06/goodlatte-bill-would-push-offshore-drilling-off-virginia-waters/.
4. United States DistrictCourt, Eastern District of Louisiana, *Hornbeck Offshore Services v. Kenneth Lee "Ken" Salazar, et al.*, Case 2:10-cv-01663-MLCF-JCW, Document 67, Filed 06/22/10.
5. Ibid.
6. "New Moratorium Applies to Any Deep-water Floating Facility," Associated Press, July 12, 2010, http://www2.tbo.com/content/2010/jul/12/122245/new-federal-moratorium-applies-any-deep-water-floa/. No longer accessible.
7. James Rosen, "Drilling Moratorium Crippling Gulf, Says Industry," Fox News, August 11, 2010, http://liveshots.blogs.foxnews.com/2010/08/11/drilling-moratorium-crippling-gulf-says-industry/.
8. Ibid.
9. John M. Broder, "White House Lifts Ban on Deepwater Drilling," *New York Times*, October 12, 2010.
10. "The Administration Is Slowly Reissuing Offshore Drilling Permits," Institute for Energy Research, March 23, 2011, http://www.instituteforenergyresearch.org/2011/03/23/the-obama-administration-is-slowly-reissuing-offshore-drilling-permits/.
11. Ibid.
12. Ibid.
13. http://www.eia.gov/forecasts/aeo/pdf/0383er%282011%29.pdf (page no longer available); http://www.redstate.com/vladimir/2011/03/24/obamasalazar-moratorium-has-crippled-domestic-oil-production/.
14. "Shell Will Not Drill Offshore Alaska in 2011, CEO Says," *Offshore Energy Today*, February 3, 2011, http://www.offshoreenergytoday.com/shell-will-not-drill-offshore-alaska-in-2011-ceo-says/. No longer accessible.
15. Outer Continental Shelf Lands Act, Section 4 (3).
16. Ibid., Section 2 (f) (4).
17. "West Coast Senators Introduce Bill to Protect Pacific Coast from New Offshore Drilling," press release, Office of Senator Patty Murray, January 25, 2011.
18. A figure of 11 million U.S. gallons is a commonly accepted estimate of the spill's volume and has been used by the State of Alaska's *Exxon Valdez* Oil Spill Trustee Council and the National Oceanic and Atmospheric Administration.
19. OPEC Statute, Chapter 2, Article 7, C.
20. Gonzalo Ortiz, "Seven Foreign Oil Companies to Pull Out," IPSNews, January 25, 2011, http://ipsnews.net/news.asp?idnews=54235.
21. *Iran Investment Monthly* 4, no. 50, November, 2010.
22. Jad Mouawad, "Exxon Mobil Delivers Record Profit," *New York Times*, February 1, 2008.
23. Editorial, "What Is a 'Windfall' Profit, *Wall Street Journal*, August 4, 2008.
24. Center for Responsive Politics, Diane Feinstein (D-CA), 2009, http://www.opensecrets.org/pfds/CIDsummary.php?CID=N00007364&year=2009. No longer accessible.
25. David M. Herszenhorn, "Senators Sharply Question Oil Officials," *New York Times*, May 22, 2008.
26. Ibid.
27. Barack Obama, presidential campaign speech, July 30, 2008.

28. The companies under investigation were ExxonMobil, Royal Dutch Shell, BP, Chevron, ConocoPhillips, Marathon, Amerada Hess, Occidental, and Murphy. Their total net revenue was $1.6 Trillion, with a net income (profit) of $127.9 Billion. CRS Report for Congress, "Oil Industry Profit Review 2007," April 4, 2008.
29. Christopher Helman, "The Other Face of Saudi Aramco," *Forbes*, August 11, 2008.
30. Ibid.
31. Determined by evaluating Exxon Mobil Corporation's 2007 annual shareholder statements.
32. Determined by evaluating Exxon Mobil Corporation's 2008 annual shareholder statements.
33. Tax Foundation, Oil Profits and Taxes, 1981–2006, April 2, 2008, http://www.taxfoundation.org/taxdata/show/23070.html.
34. Andrew Leonard, "Jimmy Carter—the Peak Oil President," *Salon.com*, August 6, 2008, http://www.salon.com/2008/08/06/jimmy_carter_peak_oil/; Max Schulz, "Running Out of Oil? History, Technology and Abundance, Manhattan Institute for Policy Research, March 13, 2006, http://www.manhattan-institute.org/html/miarticle.htm?id=4085.
35. Remarks by Abdallah S. Jum'ah, President & Chief Executive Officer, Saudi Aramco, Third OPEC International Seminar, Vienna, September 12–13, 2006, "The Impact of Upstream Technological Advances on Future Oil Supply," http://www.opec.org/opec_web/static_files_project/media/downloads/press_room/Abdallah_Jumah.pdf.
36. Morris A. Adelman, "Modeling World Oil Supply," *Energy Journal,* January 1, 1993.
37. Crude Oil Production, "Petroleum and Other Liquids," U.S. Energy Information Administration, April 3, 2011, http://www.eia.doe.gov/dnav/pet/pet_crd_crpdn_adc_mbbl_m.htm.
38. "Crude Oil and Total Petroleum Imports Top 15 Countries," U.S. Energy Information Administration, March 30, 2011, http://www.eia.doe.gov/pub/oil_gas/petroleum/data_publications/company_level_imports/current/import.html.
39. "U.S. Backs $1 Billion Loan to Mexico for Oil Drilling Despite Obama Moratorium," Fox News, September 11, 2010, http://www.foxnews.com/politics/2010/09/11/backs-b-loan-mexico-oil-drilling-despite-obama-moratorium/.
40. US Energy Information Administration, "World Proved Reserves of Oil and Natural Gas, Most Recent Estimates," March 3, 2009.
41. US Energy Information Administration, "Frequently Asked Questions," http://tonto.eia.doe.gov/tools/faqs/faq.cfm?id=33&t=6, accessed April 3, 2011.

CHAPTER 9

1. Institute of Social and Economic Research at the University of Alaska, "Research Matters No. 52," March 10, 2011, http://www.iser.uaa.alaska.edu/Home/researchmatters_2009.html. No longer accessible.
2. James Glover, *A Wilderness Original: The Life of Bob Marshall* (Seattle: The Mountaineers, 149.
3. From the Wilderness Society's 2009 Federal tax return, http://wilderness.org/files/TWS-990-FY2010.pdf.
4. Ibid.
5. Aldo Leopold, "Erosion as a Menace to the Social and Economic Feature of the Southwest," *Journal of Forestry* 44, no. 9 (September 1946): 627–33.
6. Aldo Leopold, *A Sand County Almanac* (Oxford University Press, 1949), 295.
7. http://wilderness.org/about-us/olaus-murie.
8. .01 percent.
9. U.S. Department of the Interior, Bureau of Land Management, "Oil Shale," Rock Springs, Wyoming Field Office, updated January 13, 2011, http://www.blm.gov/wy/st/en/field_offices/Rock_Springs/minerals/oil_shale.html.

10. Tashia Tucker, "The Truth About Oil Shale," Wilderness Society blog entry, May 15, 2009, http://wilderness.org/content/truth-about-oil-shale.
11. "Government Ignores Process: Opens 1.9 Million Acres to Oil Shale Development," press release, Wilderness Society, October 7, 2007.
12. "Obama Admin to Pursue Oil Shale Development," *Sustainable Business News*, February 26, 2009, http://www.sustainablebusiness.com/index.cfm/go/news.display/id/17751.
13. "Secretary of Interior Salazar Revisits Oil Shale Rules," Channel 4, Denver, CO, February 15, 2011, http://denver.cbslocal.com/2011/02/15/secretary-of-interior-salazar-revisits-oil-shale-rules/.
14. Department of Energy Office of Petroleum Reserves—Strategic Unconventional Fuels, "Fact Sheet: US Oil Shale Economics, 2006," http://www.fossil.energy.gov/programs/reserves/npr/Oil_Shale_Economics_Fact_Sheet1.pdf.
15. James T. Bartis et al., "Oil Shale Development in the United States," Rand Corporation, 2005, http://www.rand.org/pubs/monographs/2005/RAND_MG414.pdf.
16. Ibid.
17. Department of Energy Office of Petroleum Reserves-Strategic Unconventional Fuels, "Fact Sheet: US Oil Shale Economics, 2006."
18. The Rand Corporation's estimate refers to individuals employed by the plant operator at or near the plant site and includes plant employees, mine workers; and technical, management, and administrative staff.
19. Federal corporate tax rates top out at approximately 35 percent of profits, and state tax rates range around 5 percent. Oil production royalty payments to the federal government generally run between one-eighth and one-sixth of total revenues.
20. Christian Romer, Chair-Nominee-Designate, Council of Economic Advisors, "The Job Impact of the American Recovery and Investment Plan," January 9, 2009.
21. Petro Strategies, Inc., "People Who Work in the Oil and Gas Industry," Updated March 23, 2011, data compiled from US Bureau of Labor Statistics annual "Occupational Employment Statistics Survey by Occupation," May 2010, http://www.petrostrategies.org/Learning_Center/people_who_work_in_the_oil_and_gas_industry.htm.
22. United States Census Bureau, Table 1047: Retail Trade—Establishments, Employees, and Payroll, 2006–2007, http://www.census.gov/compendia/statab/2011/tables/11s1047.pdf.
23. U.S. Census Bureau, Table 917. Utilities—Establishments, Revenue, Payroll, and Employees by Kind of Business: 2007, http://www.census.gov/compendia/statab/2011/tables/11s0917.pdf.
24. US Energy Information Administration, "Crude Oil and Total Imports Top 15 Countries."
25. Gadi Adelman and Joy Brighton, "Are We Financing Our Own Demise?" Family Security Matters, March 2, 2010, http://www.familysecuritymatters.org/publications/id.5632/pub_detail.asp.

CHAPTER 10

1. Barack Obama, during a meeting with the Editorial Board of the *Keene Sentinel* newspaper in New Hampshire, on November 25, 2007, and reported in "Nuclear Power a Thorny Issue for Candidates," National Public Radio *Morning Edition*, July 21, 2008, by David Kestenbaum. Available on youtube.com: "Sen. Barack Obama on Nuclear Power from SentinelSource.com."
2. James Hylko, Robert Peltier, "The U.S. Spent Nuclear Fuel Policy: Road to Nowhere," *Power Magazine*, May 1, 2010, http://www.powermag.com/issues/cover_stories/The-U-S-Spent-Nuclear-Fuel-Policy-Road-to-Nowhere_2651.html.
3. "Secretary Chu Announces Blue Ribbon Commission on America's Nuclear Future," press release, US Department of Energy, January 29, 2010, http://www.energy.gov/news/8584.

htm. No longer accessible.
4. "Ending Yucca Mountain," *Las Vegas Sun* editorial, Tuesday, February 15, 2011.
5. Roberta Rampton, "Rush to Kill Nuclear Dump Sets Back Waste Plans," Reuters, May 10, 2011.
6. "An Ecologist's Perspective on Nuclear Power," *Federation of American Scientists Public Issue Report,* May/June 1978.
7. As of August 1, 2011.
8. International Atomic Energy Association, Power Reactor Information System, "Latest News Related to PRIS and the Status of Nuclear Power Plants," http://www.iaea.org/programmes/a2/index.html.
9. Cited in remarks made at the Department of Energy Nuclear Energy Summit, Washington, D.C., by Admiral Frank L. Bowman, USN (Retired), President and CEO, Nuclear Energy Institute, October 8, 2008.
10. "CPSC Releases New Report on Residential Fires: Latest Data Show Record Number of Fatalities from Candle Fires," United States Consumer Product Safety Commission, Press Release, July 21, 2001.
11. Daniel Lovering, "The Machines That Cleaned Up Three Mile Island," *Scientific American,* March 27, 2009.
12. "The TMI 2 Accident: Its Impact, Its Lessons," Nuclear Energy Institute, August 2010, http://www.nei.org/resourcesandstats/documentlibrary/safetyandsecurity/factsheet/tmi2accidentimpactlessons/?page=1.
13. "What Really Happened at Three Mile Island," Greenpeace Community Blog, July 16, 2009, http://members.greenpeace.org/blog/blackbox/2009/07/16/blackbox.
14. "The TMI 2 Accident: Its Impact, Its Lessons."
15. United States Nuclear Regulatory Commission, status report for Three Mile Island Reactor 2, May 9, 2011, http://www.nrc.gov/info-finder/decommissioning/power-reactor/three-mile-island-unit-2.html.
16. "Backgrounder on Chernobyl Nuclear Power Plant Accident," United States Nuclear Regulatory Commission, updated February 4, 2011, http://www.nrc.gov/reading-rm/doc-collections/fact-sheets/chernobyl-bg.html.
17. Ibid.
18. Nuclear Energy Institute, "Frequently Asked Questions: Japanese Nuclear Energy Situation," updated May 23, 2011.
19. John Boice, speaking before the House Committee on Science, Space and Technology's Energy and Environment and Investigations and Oversight Committees, May 13, 2011.
20. Ibid.
21. http://www.ans.org/pi/resources/dosechart/docs/dosechart.pdf.
22. Because tobacco is grown in soil fertilized by phosphorous-enriched materials, which are radioactive, the cigarette smoke is considered to contain a measure of radioactivity.
23. Julia Layton, "Do Backscatter X-ray Systems Pose a Risk to Frequent Fliers?" How Stuff Works, February 27, 2007, http://science.howstuffworks.com/innovation/science-questions/backscatter.htm.
24. From the National Institute of Environmental Health Sciences "Kids Page," NIEHS is a division of the National Institutes of Health, overseen by the U.S. Department of Health and Human Services, http://kids.niehs.nih.gov/recycle.htm.
25. Ibid.
26. Robert Peltier, "U.S. Spent Nuclear Fuel Policy: Road to Nowhere [Part V: Lessons]," *Master Resource,* July 13, 2010, http://www.masterresource.org/2010/07/spent-nuke-fuel-policy-5/.
27. "Announcing a Series of Policy Initiatives on Nuclear Energy," Pub. Papers 903, October 8, 1981, in *Allied-General Nuclear Services v. United States.*
28. Dave Berlin, "Nuclear Waste Storage," Citizens Against Government Waste, posted September

23, 2004, http://membership.cagw.org/site/PageServer?pagename=issues_environment_policystatement.

29. Ibid.
30. Joseph Mann, "Nuclear Power Given Green Light," *Sun Sentinel*, May 4, 2005.
31. Senator Lamar Alexander, "Blueprint for 100 New Nuclear Power Plants in 20 Years," July 13, 2009.
32. Jason Morgan, "Comparing Energy Costs of Nuclear, Coal, Gas, Wind and Solar," *Nuclear Fissionary*, April 2, 2010, http://nuclearfissionary.com/2010/04/02/comparing-energy-costs-of-nuclear-coal-gas-wind-and-solar/.
33. I purposely did not include the prices of solar and wind, as these are not baseload power sources.
34. "Skip Bowman Builds on Nuclear Promise," Nuclear Energy Institute, April 15, 2008, http://neinuclearnotes.blogspot.com/2008/04/skip-bowman-builds-on-nuclear-promise.html.
35. Peter Alpern, "US Cedes Capability for Largest Nuclear Forgings," Industry Week, June 24, 2009, http://www.industryweek.com/articles/u-s-_cedes_capability_for_largest_nuclear_forgings_19453.aspx?ShowAll=1.
36. "Skip Bowman Builds on Nuclear Promise."

CHAPTER 11

1. "Water Supply and Use in the United States," US Environmental Protection Agency, WaterSense project, June 2008, http://www.epa.gov/watersense/docs/ws_supply508.pdf.
2. "Dams—Impacts, New Developments, Dams and Rivers: A Primer on the Downstream Effects of Dams—Water, World, Projects, Construction, River, and Bank," Library Index, http://www.libraryindex.com/pages/3227/Dams.html#ixzz1OdctRk7x.
3. According to the US Geological Survey, http://ga.water.usgs.gov/edu/wuhy.html.
4. Ibid.
5. U.S. Department of Energy, Partnership with the Environment, "Hydropower," retrieved June 19, 2011, http://hydropower.inel.gov/hydrofacts/pdfs/01-ga50627-01-brochure.pdf, retrieved June 19, 2011.
6. National Energy Education Development Project, "Hydropower," from Secondary Energy Infobook, http://www.need.org/needpdf/infobook_activities/SecInfo/HydroS.pdf.
7. Ibid.
8. Associated Press, "Hydroelectric Projects Face Forbidding Permit Costs," *Providence Journal*, December 17, 2008.
9. Paul Ehrlich and Gretchen C. Daily, "Population, Sustainability, and Earth's Carrying Capacity," Bioscience, 1992. Emphasis added.
10. Ibid. Emphasis added.
11. John Holdren, Gretchen C. Daily, and Paul Ehrlich, "The Meaning of Sustainability: Biogeophysical Apsects," distributed for the United Nations University by the World Bank Washington, D.C., 1995.
12. Joseph A. Tainter, "Complexity, Problem Solving, and Sustainable Societies," published in *Getting Down to Earth: Practical Applications of Ecological Economics*, Island Press, 1996.
13. Joseph A. Tainter, *The Collapse of Complex Societies* (Cambridge: Cambridge University Press, 1988), introduction.
14. Isaac Asimov in an interview with Bill Moyers on *World of Ideas*, Corporation for Public Broadcasting, 1988.
15. Gore, *Earth in the Balance* (Plume Books, 1993), 273. (I purchased the paperback; the original hardcover edition was published in 1992; see chapter 6, note 18).
16. According to Water History, http://www.waterhistory.org/histories/dominy/.

17. Ed Marston, "Floyd Dominy: An Encounter With the West's Undaunted Dam-builder," High Country News, August 28, 2000.
18. Colorado River Compact, 45 Stat. 1057, Congressional Record, 70th Cong, 2d Sess. At 324–325, Article I.
19. Donald Pisani, "A Tale of Two Commissioners: Frank H. Newell and Floyd Dominy," presented at the History of Reclamation symposium, Las Vegas, NV, June 18, 2002.
20. Ibid.
21. Bettina Boxall, "Dominy Dies at 100; Federal Water Official Who Oversaw Major Dam Projects, *LA Times*, April 28, 2010, http://www.latimes.com/news/science/environment/la-me-floyd-dominy-20100428,0,6956153.story?track=rss&utm_source=feedburner&utm_medium=feed&utm_campaign=Feed%3A+latimes%2Fnews%2Fscience%2Fenvironment+%28L.A.+Times+-+Environment%29.
22. According to Desert USA, "Colorado River Basin," http://www.desertusa.com/colorado/coloriv/du_coloriv.html, accessed June 20, 2011.
23. Trinity River Basin Fish and Wildlife Restoration Act, Public Law 98–541 (page 98 Stat. 2721), enacted October 24, 1984.
24. Remarks of Bruce Babbitt, Secretary, U.S. Department of the Interior, Ecological Society of America Annual Meeting, Baltimore, Maryland, August 4, 1998.
25. Trinity River Basin Fish and Wildlife Management Reauthorization Act of 1995, Public Law 104–103 (page 110 Stat. 1338), enacted May 15, 1996.
26. The Secretary of the Interior authorized development of the Klamath Irrigation Project on May 15, 1905, under provisions of the Reclamation Act of 1902 (32 Stat. 388).
27. Klamath River Expert Panel, Final Report, Scientific Assessment of Two Dam Removal Alternatives on Chinook Salmon, June 13, 2011, iii.
28. Ibid., 10.
29. Ibid., 28.
30. Ibid., 14.
31. Ibid., footnote, i.
32. PacificCorp says the four hydroelectric plants provide electricity to 150,000 households, so I am assuming the national average of four people per household.
33. "Klamath Dams," Klamath Riverkeeper, http://www.klamathriver.org/Pacificorps-Dams.html, retrieved June 20, 2011.
34. Dictionary of American History, 2003, s.v., "Hydroelectric power," http://www.encyclopedia.com/topic/hydroelectric_power.aspx.
35. Ibid.
36. Hydropower, U.S. Department of Energy, http://www.energy.gov/energysources/hydropower.htm.
37. Navigant Consulting, "Job Creation Opportunities in Hydropower," presented to the National Hydropower Association, September 20, 2009.
38. The Nature Conservancy, Consolidated Financial Statements for the years ended June 30, 2010, and 2009, http://www.nature.org/aboutus/ouraccountability/annualreport/fs_fy2010.pdf.
39. The Nature Conservancy, "Reducing the Ecological Impact of Dams," January 31, 2011, http://www.nature.org/ourinitiatives/habitats/riverslakes/reducing-the-ecological-impact-of-dams.xml.
40. Figures based on the most recent statement available as of July 2011, as found in the 2008 audited statement, http://www.greenpeace.org/usa/Global/usa/binaries/2010/3/2008-inc-audit.pdf. No longer accessible.
41. "Don't Build Dams Everywhere, Expert Warns," Greenpeace.org, August 25, 2005, http://members.greenpeace.org/blog/odin2/2005/08/28/don_t_build_dams_everywhere_expert_warns.
42. "Oppose Carlsbad Ocean Desalination," Surfrider Foundation, http://www.surfrider.org/

campaigns/entry/oppose-carlsbad-ocean-desalination.
43. http://sandiego.surfrider.org/kill-your-lawn.
44. "Desalination: Is It Worth the Salt? A Primer on Brackish and Seawater Desalination," by the Lone Star Chapter of the Sierra Club, December 2008.
45. Paul and Anne Ehrlich and John Holdren, *Human Ecology Problems and Solutions* (San Francisco: WH Freeman and Company, 1973), 279.

CHAPTER 12

1. The Great Outdoors Initiative is reproduced in the appendix.
2. Barack Obama, "Presidential Memorandum—America's Great Outdoors," White House, April 16, 2010.
3. From a document I discovered on the Department of Interior website, titled, "Great Outdoors America—A 21st Century Treasured Landscapes Agenda." No date of publication is presented. Accessed August 15, 2011, http://www.doi.gov/whatwedo/outdoors/upload/PDFGreat_Outdoors_America.pdf.
4. U.S. Department of the Interior, "Secretary Salazar Establishes New Directorate for National Landscape Conservation System," press release, November 15, 2010.
5. This document can also be found in the appendix. The markings on the paper were made by a federal employee.
6. "UN Agency Calls for Inclusion of Farming in Talks on New Climate Change Treaty," United Nations News Centre, April 2, 2009.
7. Federal Land Policy Management Act, Section 102(a)(1).
8. Ibid., Section 205.
9. Ibid., Section 102(a)(2).
10. Ibid., Section 102(a)(7).
11. Ibid., Section 102(a)(8).
12. Ibid., Section 102(a)(11).
13. Federal Land Planning and Management Act, Section 205.
14. "Habitat," Section D., Land (Agenda Item 10 (d)), Preamble, 2. United Nations Conference on Human Settlements, Vancouver, May 31–June 11, 1976.
15. Holly Swanson, *Set up & Sold Out* (White City, OR: C.I.N Publishing, 1998), 181.
16. Wildlands Network, "Creating Landscapes for Life," http://wildlandsnetwork.org/what-we-do/wildways.
17. Ibid.
18. Wildlands Network, "Eastern Wildway," http://www.wildlandsnetwork.org/wildways/eastern-wildway.
19. Ibid., "Western Wildway," http://www.wildlandsnetwork.org/wildways/western-wildway.
20. Ibid., "Pacific Wildway," http://www.wildlandsnetwork.org/wildways/pacific-wildway.
21. Ibid., "Western Wildway."
22. Multicultural Environmental Leadership Development Initiative, biography, Don Chen, accessed August 19, 2011, http://meldi.snre.umich.edu/node/12431.
23. Doris Duke Foundation, Grants Awarded and Conservation Achievements, accessed from website, August 19, 2011, http://www.ddcf.org/Environment/.
24. Regional Plan Association, "America 2050: A Prospectus," New York, September 2006.
25. Ibid.
26. California SB 2, signed April 12, 2011.
27. California Air Resources Board, Resolution 10–42, December 16, 2010.
28. http://www.arb.ca.gov/msprog/onrdiesel/documents/multirule.pdf.

29. Effective January 1, 2008.
30. Effective October 20, 2010.
31. 375, Section 1, (a). Approved by Governor Arnold Schwarzenegger September 30, 2008.
32. Altmaier and others, "Make It Work: Implementing Senate Bill 375," Center for a Sustainable California, Institute of Urban and Regional Development, University of California, Berkeley, October 4, 2009, 22.
33. Ibid., 41.
34. SB 375, Section 4, (E), (i). Approved by Governor Arnold Schwarzenegger September 30, 2008.
35. Altmaier and others, "Make It Work," 57.
36. Ibid., 58.
37. S. 1619, Section 2 (14), August 6, 2009.
38. Ibid., Section 3.
39. Included in the appendix.

AFTERWORD

1. Alexander Hamilton, John Jay, James Madison, *The Federalist on the New Constitution* (Philadelphia: Benjamin Warner, 1818), p. 53, #10, James Madison.
2. B. J.Lossing, "Signers of the Declaration of Independence," (New York: Cooledge, 1848), Introduction.
3. Harry Clinton, Mary Wolcott Green, *The Pioneer Mothers of America*, Volume 3 (New York: G.P. Putnam's Sons, 1912), p. 121.

APPENDIX

1. Sir Arthur G. Tansley, "The Use and Abuse of Vegetational Concepts and Terms, Ecology, 16, no. 3 (July 1935), pg. 299, 303-304.
2. Charles Elton, *The Ecology of Invasions by Animals and Plants*, London: Methuen and Co. (1958), 137–42.
3. Rachel Carson, *Silent Spring*, (Houghton Mifflin, New York, 1962) 155.
4. Paul Ehrlich, *The Population Bomb*, (Ballantine Books, New York, 1968) 1.
5. http://www.mauricestrong.net/2008072115/strong-biography.html. Accessed January 3, 2011
6. "Habitat," Section D., Land (Agenda Item 10 (d)), Preamble, 1. United Nations Conference on Human Settlements, Vancouver, May 31–June 11, 1976.
7. William K. Reilly, "The Road from Rio," *EPA Journal*, September/October 1992, posted on the Environmental Protection Agency Website, retrieved September 11, 2011, http://www.epa.gov/aboutepa/history/topics/summit/01.html.
8. President's Council on Sustainable Development, Revised Charter, Scope of Activities, April 25, 1997.

INDEX